UMAP MODULES

Consortium for Mathematics
and Its Applications, Inc.
271 Lincoln Street, Suite No. 4
Lexington, MA 02173

This material was prepared with the partial support of National Science Foundation Grants No. SED80-07731 and SPE-8304192. Recommendations expressed are those of the authors and do not necessarily reflect the views of the NSF or the copyright holder.

ISBN 0-912843-07-1
Printed in USA

Table of Contents

Introduction VII

Numerical Listing of UMAP Modules 1984 XI

Population Dynamics of Governmental Bureaus (Unit 494)
Thomas W. Casstevens 1

Evaluating and Analyzing Probabilistic Forecasts (Unit 572)
J. Frank Yates 27

Probability in a Contested Election (Unit 634)
Dennis C. Gilliland 73

Cassette Tapes: Predicting Recording Time (Unit 641)
Arnold J. Insel 87

The Unit of Analysis and the Independence of Observations (Unit 651)
Thomas R. Knapp 103

Spacecraft Attitude, Rotations and Quaternions (Unit 652)
Dennis Pence 129

The Ricker Salmon Model (Unit 653)
Raymond N. Greenwell 173

Controlling the Effects of Interruption (Unit 657)
Jo Anne Growney 197

Windchill (Unit 658)
William Bosch and L.G. Cobb 235

The Mathematics of Focusing a Camera (Unit 659)
Raymond N. Greenwell 253

Where Are the Russian and Chinese Missiles Coming From? (Unit 661)
Sidney H. Kung 277

Computer Implementation of Matrix Computations and Row Transformations on Matrices with Application to Solving Systems of Linear Equations (Unit 662)
Wendell Motter 295

Introduction

The instructional modules in this volume were developed by the Undergraduate Mathematics and Its Applications Project (UMAP). UMAP has been funded by grants from the National Science Foundation to Education Development Center, Inc. (1976-February 1983) and to the Consortium for Mathematics and Its Applications, Inc. (February 1983-February 1985). Project UMAP develops and disseminates instructional modules and expository monographs in the mathematical sciences and their applications for undergraduate students and instructors.

UMAP modules are self-contained (except for stated prerequisites), lesson-length, instructional units from which undergraduate students learn professional applications of mathematics and statistics to such fields as biomathematics, economics, American politics, numerical methods, computer science, earth science, social sciences, and psychology. The modules are written and reviewed by classroom instructors in colleges and high schools throughout the United States and abroad. In addition, a number of people from industry are involved in the development of instructional modules.

In addition to the annual collection of UMAP modules, COMAP also distributes individual UMAP instructional modules, *The UMAP Journal*, and the UMAP expository monograph series. Thousands of instructors and students have shared their reactions to the use of these instructional materials in the classroom. Comments and suggestions for changes are incorporated as part of the development and improvement of the materials.

The substance and momentum of the UMAP Project comes from the thousands of individuals involved in the development and use of UMAP's instructional materials. In order to capture this momentum and succeed beyond the period of federal funding, we established the Consortium for Mathematics and Its Applications (COMAP) as a non-profit organization. COMAP is committed to the improvement of mathematics education, to the continuation of the development and dissemination of instructional materials, and to fostering and enlarging the network of people involved in the development and use of materials. COMAP deals with science and mathematics education in secondary schools, teacher training, continuing education, and industrial and government training programs.

Incorporated in 1980, COMAP is governed by a Board of Trustees: President W. T. Martin (M.I.T.), Treasurer George Springer (Indiana University), Vice President/Clerk William F. Lucas (Cornell University), Joseph Malkevitch (York College, CUNY), and Lynn A. Steen (St. Olaf College).

Instructional programs are guided by the Consortium Council, whose members are variously elected by the broad COMAP membership, or appointed by cooperating organizations (Mathematical Association of America, Society for Industrial and Applied Mathematics, National Council of Teachers of Mathematics). The 1985 Consortium Council is chaired by Joseph Malkevitch (York College, CUNY), and its members are:

This collection of modules represents the spirit and ability of scores of volunteer authors, reviewers, and field-testers (both instructors and students). The modules also present various fields of application as well as different levels of mathematics. COMAP is very interested in receiving information on the use of modules in various settings. We invite you to contact us:

COMAP, Inc.
271 Lincoln Street, Suite 4
Lexington, MA 02173

NUMERICAL LISTING OF UMAP MODULES 1984

Unit		Page
494	Population Dynamics of Governmental Bureaus Thomas W. Casstrvens	1
572	Evaluating and Analyzing Probabilistic Forecasts J. Frank Yates	27
634	Probability in a Contested Election Dennis C. Gilliland	73
641	Cassette Tapes: Predicting Recording Time Arnold J. Insel	87
651	The Unit of Analysis and the Independence of Observations Thomas R. Knapp	103
652	Spacecraft Attitude, Rotations and Quaternions Dennis Pace	129
653	The Ricker Salmon Model Raymond N. Greenwell	173
657	Controlling the Effects of Interruption Jo Anne Growney	197
658	Windchill William Bosch and L.G. Cobb	235

659 The Mathematics of Focusing a Camera — 253
Raymond N. Greenwell

661 Where are the Russian and Chinese Missiles Coming From? — 277
Sidney H. Kung

662 Computer Implementation of Matrix Computations and Row Transformations on Matrices with Application to Solving Systems of Linear Equations — 295
Wendell Motter

UMAP

Modules in Undergraduate Mathematics and its Applications

Module 494

Population Dynamics of Governmental Bureaus

Thomas W. Casstevens

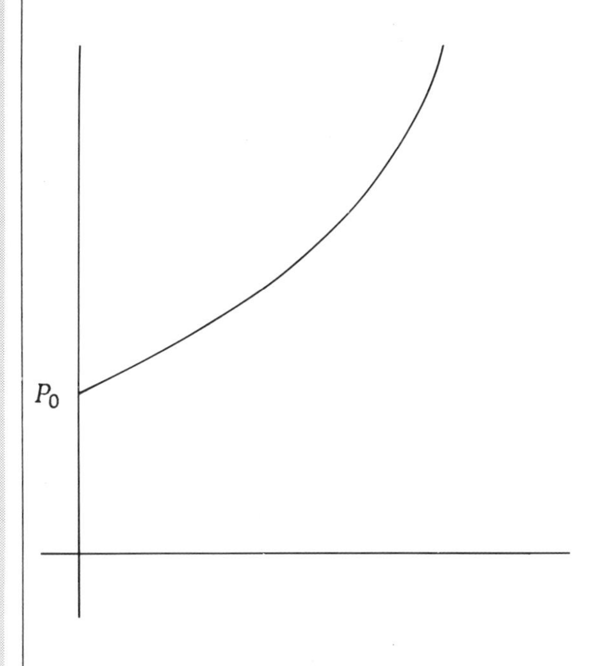

Published in cooperation with the Society for Industrial and Applied Mathematics, the Mathematical Association of America, the National Council of Teachers of Mathematics, the American Mathematical Association of Two-Year Colleges, and The Institute of Management Sciences.

INTERMODULAR DESCRIPTION SHEET: UNIT 494

TITLE: POPULATION DYNAMICS OF GOVERNMENTAL BUREAUS

AUTHOR: Thomas W. Casstevens
Department of Political Science
Oakland University
Rochester, MI 48063

MATH FIELD: Calculus

APPLICATION FIELD: Political Science

TARGET AUDIENCE: Students in a second semester calculus course

ABSTRACT: This module presents, calibrates, and tests a mathematical model of the birth and death of bureaucratic units—briefly, bureaus—in the executive departments (outside the defense and postal services) of the national government of the United States. The model formalizes the popular notion that bureaus proliferate like rabbits, i.e., exponentially.

PREREQUISITES: Integration of exponential and logarithmic functions, elementary probability.

The UMAP Journal, Vol. V, No. 2, 1984

Population Dynamics of Governmental Bureaus

Thomas W. Casstevens
Department of Political Science
Oakland University
Rochester, MI 48063

Table of Contents

1. INTRODUCTION 1
2. DATA 2
3. MODEL 4
4. GROWTH = BIRTHS − DEATHS 5
5. ESTIMATION 6
6. BIRTHS 7
7. DEATHS 10
8. INTERLUDE 12
9. AGES 12
10. PROBABILITY 14
11. MEMORY 15
12. LOGISTIC? 17
13. CONCLUSION 19
14. ANSWERS TO EXERCISES 20

Modules and Monographs in Undergraduate Mathematics and its Applications Project (UMAP)

The goal of UMAP was to develop, through a community of users and developers, a system of instructional modules in undergraduate mathematics and its applications to be used to supplement existing courses and from which complete courses may eventually be built.

The Project was guided by a National Advisory Board of mathematicians, scientists, and educators. UMAP was funded by a grant from the National Science Foundation and is now supported by the Consortium for Mathematics and Its Applications, Inc. (COMAP), a nonprofit corporation engaged in research and development in mathematics education.

The Project would like to thank Gilbert Lewis of Michigan Technological University, and J.M. Elkins of Sweet Briar College, for their reviews, and all others who assisted in the production of this unit.

Early reports of this research were presented at the Western Political Science Association Annual Meeting (1979) and published in *Virginia Municipal Review* (Vol. 57, No. 6, June 1979, p. 2) as well as *Behavioral Science* (Vol. 25, No. 2, March 1980, pp. 161-165).

1. Introduction

This module presents, calibrates, and tests a mathematical model of the birth and death of bureaucratic units—briefly, bureaus—in the executive departments (outside the defense and postal services) of the national government of the United States. The executive departments do not include the so-called independent agencies.

The model formalizes the popular prejudice that bureaus proliferate like rabbits. A classical model of growth is the exponential model,

$$P(t) = P_0 e^{kt}, \tag{1}$$

where $e = 2.718\ldots$, k is the rate of growth (a positive constant), t is time in suitable units, $P(t)$ is the number (of bureaus) at time t, and P_0 is the number (of bureaus) at time zero. Figure 1 displays this characteristic pattern of exponential growth graphically. The growth in the number of bureaus fits this model to a good approximation, as shown below.

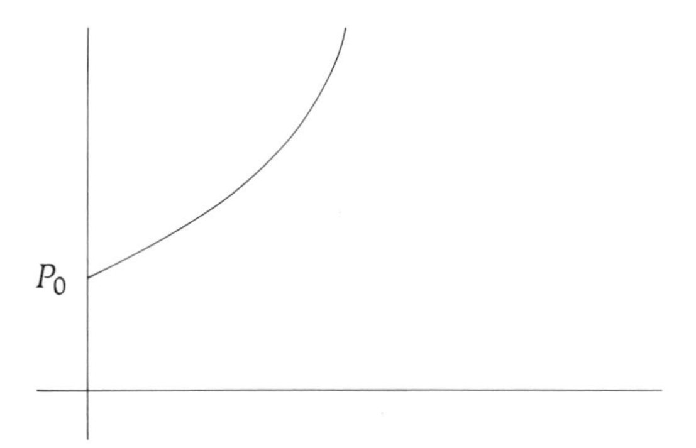

Figure 1. Exponential growth

Equation (1) is qualitative until values are estimated for k and P_0. Furthermore, aside from time, the model does not specify the causes of growth, although the rabbits metaphor suggests that bureaus beget bureaus.

Nevertheless, the model is informative. The model, as given, shows that the growth in the number of bureaus is not constrained and, consequently, suggests that the growth in the number of bureaus is not controlled; presidential protestations to the contrary notwithstanding. Reality is obscured by verbiage in politics and government!

The problem of control is not solved in this module. Equation (1), nevertheless, contributes to a formulation of the problem of control. For purposes of control, politicians need a function f of one or more variables $x_1, x_2, \ldots, x_n$ whose historic values are such that $f(x_1, x_2, \ldots, x_n)$ approximately equals $P(t)$ from 1789 to 1983. Control of growth is possible if and only if the government can manipulate the values of one or more of these variables. The function f and its variables $x_1, x_2, \ldots, x_n$ have not yet been discovered.

The problem of estimation is solved in the sequel, using data from 1789-1974. Equation (1), thereby, becomes a quantitative model that is a guide for research on the problem of control as well as a description of the history of bureaucracy in America.

2. Data

What is a bureau? The unit of a bureaucracy! This question and its answer are troublesome, despite their simplicity.

The existence of a bureau is not obvious in general. Bureaus have many names: agency, office, bureau, and so forth. These labels may—but only may—denote the existence of a bureau.

The birth or death of a bureau is also not obvious in general. Bureaus are reorganized, amalgamated, relocated, and so forth. These labels may—but only may—denote the creation or destruction of a bureau.

The government itself does not count its bureaus. This must be done by other observers. Government documents contain the relevant information, but not all documents are public documents. (What is the real organization chart of the Central Intelligence Agency?) Furthermore, even if public, the documents must be sifted with care.

Bureaus are difficult to count. Herbert Kaufman, nevertheless, undertook that task and reported the results in his monograph *Are Government Organizations Immortal?* (Washington, D.C., The Brookings Institution, 1976).

Kaufman examined all units in the executive departments, aside from the defense departments and postal services, for 1923 and 1973. (The bureaus were studied through 1973. Thus, for some measurements, the relevant years are 1923 and 1974.) The births and deaths of those bureaus were scrutinized with particular care (and difficulty).

Kaufman found that **"WHEN THE ORGANIZATIONS ALIVE IN 1973 ARE ARRANGED ACCORDING TO DATE OF BIRTH AND THE CUMULATIVE NUMBERS ALIVE IN EACH**

PRESIDENTIAL TERM ARE PLOTTED, THE CURVE PRODUCED ASSUMES AN EXPONENTIAL FORM." Kaufman was right to capitalize that sentence; it is a capital finding. (See Exercise 1 for the heuristic significance of the cumulative numbers.)

Exercise 1.
The choice of format for data is vital for the discovery of models. This choice is made by the scientist. Kaufman wisely chose a cumulative format, but this choice was not required for the 394 bureaus of 1973. Table 2 gives the numbers of those 394 that were born *by* decade n before 1973. An exponential pattern "leaps to the eye" from the graph.

(**a**) Graph the numbers of Table 2.

But the data could have been viewed in terms of the numbers of those 394 that were born *during* decade n before 1973. These numbers can be derived from Table 2.

(**b**) Derive the cited numbers from Table 2. Graph the derived numbers, using a full sheet of graph paper. Does any pattern "leap to the eye" from the graph?

The data in general are vectors of the form (t_i, o_i), where $t_0, t_1, \ldots, t_n$ is a sequence of times of observations and $o_0, o_1, \ldots, o_n$ is the (associated) sequence of number of bureaus, and where $t_0 = 0$ and $o_0 = P_0$.

The data are cumulative on deaths as well as births of bureaus: Tables 1 and 2 (below) describe the births of the bureaus that existed in 1923 or in early 1974. Table 3 (below) records the deaths of the bureaus that existed in 1923.

The data on births and deaths, as given, are not sufficient for a direct test of the exponential model of bureaucratic growth. Nevertheless, since growth is the difference between births and deaths, the data should be sufficient for an indirect test of Equation (1). This intuitive insight suggests that there exists a pay-off in a theoretical* analysis of bureaucratic growth. (*Empirical *and* mathematical.)

The strategy is to derive the model of growth from models of birth and death. Kaufman suggests an exponential equation as a birth model, and thus, mere association suggests an exponential equation as a death model. This line of attack is pursued in detail, following a preliminary calibration of Equation (1).

3. Model

The data as given are sufficient to calibrate the model of Equation (1). For this purpose, that model is accepted provisionally, as a working hypothesis.

Equation (1) has time as its independent variable. All else is constant. The value of P_0 is furnished (at least occasionally) by Kaufman. The value of e is known. But the value of k is not known (in advance).

The constant k is the growth rate, and in part, its value depends upon the unit of measurement for time. For convenience, this module uses years as units of time, following Kaufman.

Kaufman found that the number of bureaus was 175 in 1923 and 394 in 1974. Equation (1), thus, gives

$$175 = P(t) = P_0 e^{kt} \tag{2}$$

and

$$394 = P(t + 51) = P_0 e^{k(t+51)} \tag{3}$$

$$= P_0 e^{kt + 51k} = P_0 e^{kt} e^{51k},$$

so by division

$$394/175 = e^{51k}, \tag{4}$$

for these data. Equation (4) is solved by taking the natural logarithms of both sides:

$$ln\ e^{51k} = 51k = ln\ (394/175) = 0.8115649, \tag{5}$$

where ln denotes the natural logarithm. Thus, if growth is exponential, $0.8115649/51 = 0.0159$ is an estimate of the value of k.

The First Congress (1789-1791) had the task of setting up the organization of the government of the United States. Historically speaking, P_0 is the number of bureaus (aside from the defense and postal services) created during the First Congress. Equation (1), with this information and $k = 0.0159$, yields

$$175 = P_0 e^{(0.0159)(1923\text{-}1791)} \tag{6}$$

$$= P_0 e^{(0.0159)(132)}$$

$$= P_0 e^{2.0988}$$

$$= P_0 (8.156)$$

for 1923. Thus, if growth is exponential, $175/8.156 = 21.5$ is an estimate of the value of the historical P_0. (See Exercise 2.)

Exercise 2.

(**a**) Find P_0 for $394 = P_0 e^{(0.0159)(1974\text{-}1791)}$.

(**b**) Compare the text's $P_0 = 21.5$ with the result of (**a**). Comment.

A testable retrodiction (not prediction) has been deduced during the preliminary calibration of the growth model: P_0 was 21 or 22 in 1791. Kaufman, alas, did not report a value for P_0 in 1791. Thus, although possible in principle, this test of the model is not convenient in practice.

We note in passing that the estimates, $k = 0.0159$ and $P_0 = 21.5$, are based upon an absolute minimum of relevant information inasmuch as two equations in two unknowns cannot be solved with fewer than two data points. These estimates, consequently, are quite tentative.

4. Growth = Births − Deaths

The growth rate of a population is a net figure, namely, a function of a birth rate b and a mortality rate m. For simplicity, as a conjecture, let $k = b - m$. Equation (1) then becomes

$$P(t) = P_0 e^{kt} = P_0 e^{(b-m)t} \tag{7}$$

$$= P_0 e^{bt - mt} = P_0 e^{bt} e^{-mt},$$

where b and m are positive constants.

When $m = 0$, the process becomes a pure birth process, and the number of bureaus at time t is

$$P(t) = P_0 e^{bt} \tag{8}$$

because $e^{-mt} = e^{-(0)t} = e^0 = 1$. This conjecture is plausible. Tables 1 and 2, when graphed, visually suggest that the birth process is an exponential process.

When $b = 0$, the process becomes a pure death process, and the number of bureaus at time t is

$$P(t) = P_0 e^{-mt}. \tag{9}$$

This conjecture is defensible. Table 3, when graphed, suggests that the death process is approximately linear with a low mortality rate for the period 1923-1974. And a low rate of exponential decay is approximately linear for any fairly lengthy period. (See Figure 2, Exercise 3 for illustrations of this approximation.)

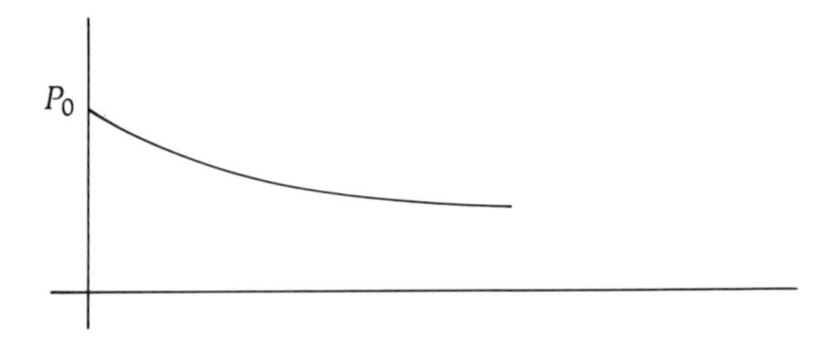

Figure 2. Exponential decay

Exercise 3.

(**a**) Calculate $P(t) = e^{-ct}$ for $c = 0.00578$ and $t = 0, 10, 20, 30, 40,$ and 50.

(**b**) Graph the results of (**a**).

(**c**) With a straight edge, draw a straight line that comes as close as possible to every point in the graph. Does this line touch every point?

Kaufman's data describe pure birth and death processes (although this is only implicit in his monograph). Equations (8) and (9), thus, can be tested directly with his data. Equation (1), consequently, can be tested indirectly since it is implied by Equations (8) and (9). (See Equation (7).)

5. Estimation

This section is a statistical interlude. Estimation methods are a central topic of applied statistics, and properly so, but this module is casual with its statistics. The justification for this casualness, aside from limits of space, is that if a mathematical model is obviously good—that is, if it describes an empirical phenomenon closely—then all sensible methods yield approximately identical estimates. Conversely, if approximately identical estimates flow from different estimation methods, then confidence is increased in a mathematical model.

Equations (1), (8), and (9) pose the same type of estimation problem for k, b, and m, respectively, because the equations have the same exponential form. For notational convenience, therefore, let the parameter c denote the positive constant k, b, or m. (Briefly, $c = k, b$, or m.) Statisticians have a variety of methods for finding a solution to this type of problem. This module uses a simple method.

A MEAN VALUE METHOD: The exact value of the parameter c is calculated for each observation and then, the mean (average) of those exact values is used as the model value. This method is quick if the number of observations is small.

The exact value of the parameter c_i for an observation o_i of time t_i is

$$c_i = |ln\ o_i - ln\ P_0| / t_i. \tag{10}$$

(See Exercise 4.) The mean of these values is easy to calculate, but as shown below, some ingenuity may be required for the selection of a time scale and P_0.

Exercise 4.

(**a**) Derive Equation (10) from Equation (1) for $c_i = k_i$.

(**b**) Derive Equation (10) from Equation (8) for $c_i = b_i$.

(**c**) Derive Equation (10) from Equation (9) for $c_i = m_i$.

6. Births

Kaufman studied the bureaus that existed in 1923 or in 1973. Experimentally speaking, the initial population P_0 is the number of bureaus that existed in 1923 or in 1973. This stipulation plays havoc with ordinary time, but it reflects the distinguished role of the initial observation in the exponential model. The value of P_0 clearly is distinguished in the equations; in practical counting all observations refer to the initial observation. The connection is plain if proportions are used; for P_0 is the denominator in all proportions.

Table 1

Age of Bureaus in 1923

Year	Age in Years ($=t$)	Number of Bureaus At Least t Years Old	$175e^{-0.0236t}$ (Model Number)
1923	0	175	175
1913	10	140	138.2
1903	20	102	109.2
1893	30	85	86.2
1883	40	70	68.1
1873	50	60	53.8
1863	60	44	42.5
1853	70	31	33.5
1843	80	23	26.5
1833	90	21	20.9
1823	100	15	16.5
1813	110	13	13.0
1803	120	11	10.3
1793	130	11	8.1

1923 EXAMPLE. There were 175 bureaus in 1923. Table 1 tabulates their births: 11 were established at least 130 years ago, ..., 44 were established at least 60 years ago, ..., and, of course, 175 were established at least 0 years ago (relative to 1923).

1923 (CONTINUED). The mean value method yields 0.0236 as the mean of the exact parameters for the data in Table 1. This mean estimate has been used to compute the theoretical numbers in Table 1. A visual comparison of the theoretical and observed numbers suggests that the model is quite good. (See Exercise 5 for computational details.)

Exercise 5.

(**a**) Find the exact parameters for the data in Table 1.

(**b**) With naive intuition, comment upon the results of (**a**).

(**c**) With normal statistics, comment upon the results of (**a**).

(**d**) With regression analysis, comment upon the results of (**a**).

Note: This type of exercise is pertinent also for Tables 2 and 3.

Table 2

Age of Bureaus in 1973

Year	Age in Years ($=t$)	Number of Bureaus At Least t Years Old	$394e^{-0.0225t}$ (Model Number)
1973	0	394	394
1963	10	281	314.6
1953	20	241	251.2
1943	30	185	200.6
1933	40	155	160.2
1923	50	148	127.9
1913	60	121	102.1
1903	70	88	81.6
1893	80	73	65.1
1883	90	60	52.0
1873	100	52	41.5
1863	110	38	33.2
1853	120	27	26.5
1843	130	19	21.1
1833	140	18	16.9
1823	150	14	13.5
1813	160	12	10.8
1803	170	10	8.6
1793	180	10	6.9

1973 EXAMPLE. There were 394 bureaus in 1973. Table 2 tabulates their births: 10 were established at least 180 years ago, ..., 60 were established at least 90 years ago, ..., and of course, 394 were established at least 0 years ago (relative to 1973).

1973 (CONTINUED). The mean value method yields 0.0225 as the mean of the exact parameters for the data in Table 2. This mean estimate has been used to compute the theoretical numbers in Table 2. A visual comparison of the theoretical and observed numbers suggests that the model is not too bad. (See Exercise 6 for computational details.)

Exercise 6.
For $b = 0.0225$, calculate the theoretical numbers in Table 2. Note: This type of exercise is pertinent also for Tables 1 and 3.

The estimates of the values of the parameters for the birth processes are roughly the same for the bureaus of 1923 and 1973. This suggests that the birth process has been rather stable historically (as conjectured). The mean of the estimates is $(0.0236 + 0.0225)/2 = 0.02305 = b$. This mean estimate is used in subsequent calculations.

7. Deaths

Kaufman studied the futures of the 175 bureaus of 1923. He traced them to extinction or through 1973. Table 3 tabulates their survival year by year.

The mean value method yields 0.00578 as the mean of the exact parameters for the data in Table 3. This mean estimate is used in subsequent calculations. (See Exercise 7 for theoretical numbers to compare with Table 3.)

Exercise 7.
For $m = 0.00578$, calculate the theoretical numbers of surviving bureaus for $t = 0, 10, 20, 30, 40,$ and 50. (*Hint:* See Exercise 3.) Compare the results with the (pertinent) observed numbers in Table 3. Comment.

Table 3

Survival of Bureaus, 1923-1973

Year	Time	Number of Bureaus That Survived From $Time_0$ to Time *t*	Year	Time	Number of Bureaus That Survived From $Time_0$ to Time *t*
1923	0	175	1949	26	157
1924	1	171	1950	27	156
1925	2	171	1951	28	156
1926	3	169	1952	29	154
1927	4	167	1953	30	149
1928	5	166	1954	31	149
1929	6	166	1955	32	148
1930	7	166	1956	33	148
1931	8	166	1957	34	148
1932	9	166	1958	35	148
1933	10	164	1959	36	148
1934	11	163	1960	37	148
1935	12	163	1961	38	148
1936	13	163	1962	39	148
1937	14	163	1963	40	148
1938	15	162	1964	41	148
1939	16	161	1965	42	148
1940	17	161	1966	43	148
1941	18	159	1967	44	148
1942	19	157	1968	45	148
1943	20	157	1969	46	148
1944	21	157	1970	47	148
1945	22	157	1971	48	148
1946	23	157	1972	49	148
1947	24	157	1973	50	148
1948	25	157			

8. Interlude

The mortality rate m and the birth rate b have now been estimated with extensive data, and finally, the growth rate k can be estimated with some confidence: $k = b - m = 0.02305 - 0.00578 = 0.01727$ should be a good approximation. This estimate exceeds the preliminary estimate by $0.01727 - 0.0159 = 0.00137$. Exponential functions are rather sensitive to changes in the values of their parameters, so this difference is not trivial for predictions of the far future or retrodictions of the distant past. For example, with $k = 0.01727$ rather than 0.0159, the mean $P_0 = 17.3$ rather than 21.5 for 1791. (See Section 3 and Exercises 2 and 8.)

Exercise 8
Estimate, with $k = 0.01727$, the historical P_0 of 1791.

(**a**) From 394 bureaus in 1974.

(**b**) From 175 bureaus in 1923.

(**c**) Compare the results of (**a**) and (**b**). Comment. (*Hint:* See Exercise 2.)

The estimates of the parameters are subject to revision, but the presentation and calibration of the model are now complete. The exponential model, however, has not yet been tested rigorously. Casual comparisons of theoretical and observed numbers have suggested that "there is something to it" as shown in Tables 1 and 2. (See Exercise 7 for Table 3.) But how much? This question is addressed below, where testable theorems are deduced from the exponential model.

9. Ages

Table 4 gives the distributions of the ages of the bureaus in 1923 and 1973. The two distributions are very similar, and this similarity corroborates the exponential model via the following theorem: If the birth rate and mortality rate are both constant over time, then the age distribution becomes not only theoretically constant but also empirically stable and the initial distribution "washes out" in the long run. This theorem is known as the Strong Ergodic Theorem of Demography. The theorem holds for any population with constant rates for births and deaths, so the distribution of ages of bureaus should be about the same in 1923 and 1973. (See W.B. Arthur, "Why a Population Converges to Stability," *American Mathematical Monthly*, Vol. 88, No. 8, (October, 1981), pp. 557-563.)

Table 4

Distribution of Ages of Bureaus in 1923 and 1973

Age	1923 Proportion		1973 Proportion	
00-09	.200	.417	.287	.389
10-19	.217		.102	
20-29	.097	.183	.142	.218
30-39	.086		.076	
40-49	.057	.148	.018	.087
50-59	.091		.069	
60-69	.074	.120	.084	.122
70-79	.046		.038	
80-89	.011	.045	.033	.053
90-99	.034		.020	
100-109	.011	.022	.036	.064
110-119	.011		.028	
120-129	.000	.063	.020	.068
130-	.063		.048	

The Strong Ergodic Theorem of Demography implies, among other things, that the median age should be constant in the long run for a given population. The classification by decade of age is too crude to permit the extraction of the median age of bureaus from Table 4. Kaufman, however, reported that the median age was 27 years for bureaus in 1923 as well as in 1973. This constancy corroborates the exponential model.

We note in passing that a stable distribution of ages should not be confused with a constant size of population: 175 ≠ 394, but the distribution of ages of bureaus is the same for 1923 and 1973.

10. Probability

The mortality model can be recast in the language of probability theory: The number of bureaus that survive to time t is $P_0 e^{-mt}$. (See Equation (9).) Consequently, the proportion that survives to time t is

$$P_0 e^{-mt} / P_0 = e^{-mt}, \tag{13}$$

and since survival and death are mutually exclusive and exhaustive,

$$p(t) = 1 - e^{-mt} \tag{14}$$

is the proportion that dies by time t. These proportions are the probabilities of survival (13) and death (14) by time t. The interpretation of proportions as probabilities is common in modeling.

Equation (14) is known as the exponential probability distribution. (See Exercise 9.) The expectation, half-life, and standard deviation of this probability distribution are expressible in terms of the mortality rate m: The expectation, which is the counterpart of the mean in discrete statistics, equals $1/m$. The half-life, which is the counterpart of the median in discrete statistics, approximates $0.693/m$. (See Exercise 10.) The standard deviation, which is the square root of the variance, equals $1/m$. These properties are worth memorizing for the exponential probability distribution.

Exercise 9.
A probability distribution (over the non-negative real numbers) is defined as any function $f(t)$ that has the following three properties:

(i) $f(0) = 0$.

(ii) If $t_2 > t_1$, then $f(t_2) \geq f(t_1)$.

(iii) $\lim_{t \to \infty} f(t) = 1$.

(a) Prove that $p(t) = 1 - e^{-mt}$ has the first property.

(b) Prove that $p(t) = 1 - e^{-mt}$ has the second property.

(c) Show that $p(t) = 1 - e^{-mt}$ has the third property.

(A rigorous proof is not required for this exercise.)

Exercise 10.

(**a**) Prove that $t = 0.693/m$ when $p(t) = 1 - e^{-mt} = 1/2$.

(**b**) Compute the half-life if $m = 0.02305$. (Note: Section 6 estimates that $b = 0.02305$.)

(**c**) Interpret the result of (**b**). (*Hint:* Section 7 reports a median age of 27 for the bureaus in 1923 and 1973.) Comment.

Section 7 suggests, as an estimate, that $m = 0.00578$. Consequently, the expectation of life for bureaus is about $1/0.00578 = 173.0$ years. And the half-life for bureaus is about $0.693/0.00578 = 119.9$ years. A citizen may be excused for confusing this with immortality. Americans, as individuals, have an expectation of life of only seventy-some years.

11. Memory

"Whether the age of an organization bears on its chances for survival is an open question," according to Kaufman. This question can now be answered.

The exponential probability distribution is memoryless. The absence of memory can be shown as follows: From time zero, a bureau has probability e^{-mt} of surviving to time t and a probability of $e^{-m(t+d)}$ of surviving to time $t + d$, where $d \geq 0$. Thus, given that a bureau has survived from time zero to time t, the probability of its surviving to time $t + d$ is

$$e^{-m(t+d)}/e^{-mt} \tag{15}$$

$$= e^{-mt-md}/e^{-mt}$$

$$= e^{-mt}e^{-md}/e^{-mt}$$

$$= e^{-md},$$

and with time t as (a new) time zero, this equation is the same as Equation (13)!

The mathematics may be obscure, but the moral is clear: Every time t is a new time zero for an extant bureau. The system does not remember how long a bureau has been in the system. The age of a bureau, thus, does not per se influence its chances for survival. This theorem can be tested directly. Table 5 gives the details for survival of bureaus, by decade of age, for

1923-1973. The agreement between model and data is obviously good.

We note in passing that the exponential probability distribution is the only continuous distribution with the memoryless property. The geometric probability distribution is its discrete counterpart. (See William Feller, *An Introduction to Probability Theory and Its Applications,* 3rd ed., New York: John Wiley and Sons, Inc., 1968, pp. 328-329.)

Table 5

Survival from 1923 to 1973 by Age of Bureau

Age in 1923 (in years)	Number in 1923	Expected Number of Survivors in 1973	Observed Number of Survivors in 1973
00-09	35	26.2	27
10-19	38	28.5	33
20-29	17	12.7	15
30-39	15	11.2	13
40-49	10	7.5	8
50-59	16	12.0	14
60-69	13	9.7	11
70-79	8	6.0	8
80-89	2	1.5	1
90-99	6	4.5	4
100-109	2	1.5	2
110-119	2	1.5	2
120-129	0	0	0
130-	11	8.2	10

Note: The expected number of survivors are ne^{-mt}, where n = number in 1923, e = 2.718 ..., m = 0.00578, and t = 50 (years).

12. Logistic?

The exponential model does not limit the number of bureaus in the executive departments of the national government of the United States. Indeed, as time goes on, the number of bureaus will continue to increase without bound, according to Equation (1).

Boundless growth seems implausible in the long run. How long is the long run? For bureaus, the long run is very long, on a human scale. Centuries are the units of time.

The number of bureaus is doubled in about 40 years, according to the exponential model. (See Exercise 11 for an exact estimate of the doubling time.) Thus, for the year 2003 (= 1923 + 80), the number of bureaus should be about (175 + 175) + (175 + 175) = 700. And for the year 2054 (= 1974 + 80), the number of bureaus should be about (394 + 394) + (394 + 394) = 1576. These figures are not absurd forecasts. The long run, accordingly, is further in the future.

Exercise 11.
When $P_0e^{kt} = 2P_0$, what is the value of t, to one decimal place, if $k = 0.01727$ (as estimated in Section 8)?

Broadly speaking, in the long run, empirical systems of exponential growth "go bust" or level off. Both possibilities should be considered for governmental bureaus.

The American bureaucracy may go bust in the long run, that is, the number of bureaus may decrease abruptly at some future time. A sharp decline of this sort might result, for example, from a revolutionary overthrow of the (then) existing government. And since centuries are the units of time, the possibility of a revolution cannot be dismissed for the long run (even in the United States).

The American bureaucracy may level off in the long run, that is, the number of bureaus may approach smoothly towards a maximum number M. A standard model for such a process is the logistic equation

$$P(t) = (MP_0)/[P_0 + (M - P_0)e^{-Mct}], \tag{16}$$

where c is a positive parameter. A logistic curve grows very rapidly to its inflection point where $P(t) = M/2$ and, then, grows more slowly towards the population's limiting value M. (See Figure 3.)

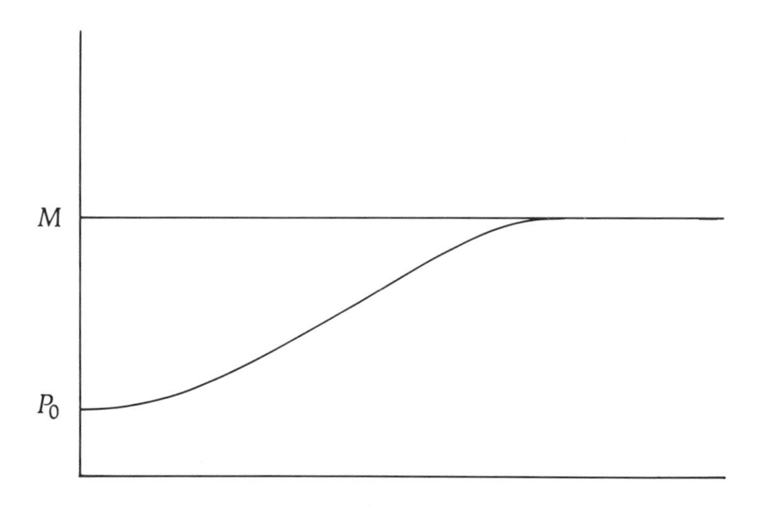

Figure 3: Logistic growth

A logistic process is difficult to distinguish from an exponential process, in practice, for values less than $M/2$. (See Exercise 12 for an illustration of this approximation.) Bureaucratic growth might be a logistic process in the United States. Perhaps the exponential model is only a convenient approximation of the historical record thus far (if, say, $M/2 > 394$).

Exercise 12.
Let $c = 0.000025$, $M = 800$, $k = 0.01727$, and $P_0 = 17$ for 1791. (The value of c is a doodle for curve fitting. See Section 12, Section 8, and Exercise 8, respectively, for the rationales for the hypothetical values of M, k, and P_0.)

(**a**) With Equation (16) compute the values of $P(t)$ for $t = 0$, 70, 132, and 183.

(**b**) With Equation (1) compute the values of $P(t)$ for $t = 0$, 70, 132, and 183.

(**c**) Compare the results of (**a**) and (**b**). Compare the results with the observed values for 1923 and early 1974. Comment.

13. Conclusion

The historical growth in the number of bureaus has been approximately exponential, outside the defense and postal services. (See Exercise 13 for a comment on the defense services.) What has driven this growth? I do not know. The form of the answer, however, is obvious: The exponential model should equal (at least approximately) a mathematical function of causal variables. (Population, budgets, employees, programs, *etcetera*, are obvious possibilities as causal variables.) The search for that function should proceed alongside an improvement in the data on the birth and death of bureaus. This subject abounds in open problems for future research.

Exercise 13.
Presumably, historically speaking, the number of units in defense has gone "boom or bust" repeatedly. Why?

14. Answers to Exercises

1. (b) 113 (in first decade before 1973), 40, 56, 30, 7, 27, 33, 15, 13, 8, 14, 11, 8, 1, 4, 2, 2, 0, 10 (in nineteenth decade before 1973). Perhaps a pattern is evident, but probably not.

2. (a) $P_0 = 21.5$
   (b) They are the same, and must be the same because $k = 0.0159$ is derived from both Equations (2) and (3).

3. (a) $P(0) = 1$, $P(10) = 0.94$, $P(20) = 0.89$, $P(30) = 0.84$, $P(40) = 0.79$, $P(50) = 0.75$.
   (c) A reasonably drawn straight line can touch every point. (Proverb: A broad pencil point ensures a good fit.)

4. (a) We have $o_i = P_0 e^{kt_i}$, where $o_i \geq P_0$. We successively obtain

$$ln\ o_i = ln\ P_0 e^{kt_i} = ln\ P_0 + ln\ e^{kt_i} = ln\ P_0 + kt_i,$$

so

$$ln\ o_i - ln\ P_0 = kt_i,$$

and

$$(ln\ o_i - ln\ P_0)/t_i = |ln\ o_i - ln\ P_0|\ /t_i = k.$$

   (b) We have $o_i = P_0 e^{bt_i}$, where $o_i \geq P_0$. This is equivalent to problem (**a**).
   (c) We have $o_i = P_0 e^{-mt_i}$, where $o_i \leq P_0$. We successively obtain

$$ln\ o_i = ln\ P_0 e^{-mt_i} = ln\ P_0 + ln\ e^{-mt_i} = ln\ P_0 - mt_i,$$

so

$$ln\ o_i - ln\ P_0 = mt_i,$$

and

$$-(ln\ o_i - ln\ P_0)/t_i = |ln\ o_i - ln\ P_0|\ /t_i = m.$$

5. (a) The exact parameters are, successively, 0.0223, 0.0270, 0.0241, 0.0229, 0.0214, 0.0230, 0.0247, 0.0254, 0.0236, 0.0246, 0.0236, 0.0231, and 0.0213.
   (b) The mean = median = mode = 0.0236. This suggests that the mean is "on target."
   (c) The standard deviation of this sample of thirteen numbers is 0.0016. 9/13 = 0.69 of the numbers are within $\pm$ 1 standard deviation of the mean. 12/13 = 0.92 of the numbers are within $\pm$ 2 standard deviations from the mean. 1/13 = 0.08 of the numbers are (slightly) beyond 2 standard deviations from the mean. The standard normal distribution has about 2/3 within $\pm$ 1 standard deviation from the mean and about 95% within $\pm$ 2 standard deviations from the mean. This suggests that the exact parameters display "normal experimental error."
   (d) The regression line (least squares straight line) is almost flat. the slope is -0.00000857 and the intercept is 0.0242. The correlation with time is very low, viz., -0.21. This suggests that the value of the parameter did not really change over time.
   Note: The results are not so nice for Tables 2 and 3.
6. The answers are given in Table 2. Note: We calculate as if this process were a death process because the numbers are decreasing, but of course, the process really is a birth process.

7. For times $t = 0, 10, 20, 30, 40$ and 50, the computing formula is $175e^{-0.00578t}$. The results of the computations are 175, 165.2, 155.9, 147.1, 138.9, and 131.1. The data are 175, 164, 157, 149, 148 and 148. The fit gets worse over time.

8. (a) $P_0 = 394e^{(0.01727)(1791\text{-}1974)} = 394e^{0.01727(-183)} = 16.7$.
   (b) $P_0 = 175e^{(0.01727)(1791\text{-}1923)} = 175e^{0.01727(-132)} = 17.9$.
   (c) The results are not identical, but they are derived from different data sets.

9. **(a)** $p(0) = 1 - e^{-m(0)} = 1 - e^0 = 1 - 1 = 0.$

(b) For $d > 0$, $1 - e^{-m(t+d)} > 1 - e^{-mt}$ because

$$1 - e^{-m(t+d)} = 1 - e^{-mt-md} = 1 - e^{-mt}e^{-md}$$

and

$$0 < e^{-md} = 1/e^{md} < 1.$$

(c) That

$$\lim_{t \to \infty} 1 - e^{-mt} = 1$$

can be seen via $1 - e^{-mt} = 1 - 1/e^{mt}$ since m is a positive constant, by increasing the value of t, the ratio $1/e^{mt}$ can be made to approach zero as closely as desired.

10. **(a)** Let $1 - e^{-mt} = 1/2$. We successively obtain $-e^{-mt} = -1/2$ and $e^{-mt} = 1/2$ and $-mt = ln(1/2) \simeq -0.693$, so $m \simeq 0.693/t$.

(b) 30.1 (years).

(c) The median age of bureaus should have been 30.1 years in 1923 as well as in 1973. There is a ten percent discrepancy between model and data.

11. $P_0e^{kt} = 2P_0$ yields $e^{kt} = 2$ and $kt = ln\ 2 \simeq 0.693$, so $t \simeq 0.693/k = 0.693/0.01727 = 40.1$ (years).

12.

Time	0	70	132	183
Logistic	17	64.7	186.6	366.1
Exponential	17	56.9	166.1	400.9
Observation	------	--------	175	394

13. The number of units "must" have increased sharply with outbreaks of war and decreased sharply with outbreaks of peace.

UMAP

Modules in Undergraduate Mathematics and its Applications

Module 572

Evaluating and Analyzing Probabilistic Forecasts

J. Frank Yates

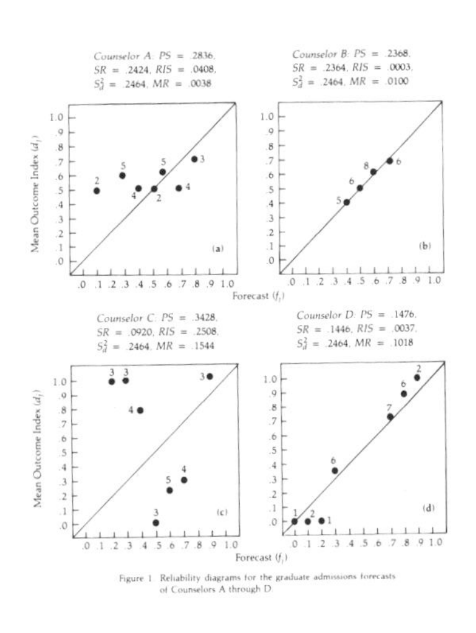

Figure 1. Reliability diagrams for the graduate admissions forecasts of Counselors A through D.

Published in cooperation with the Society for Industrial and Applied Mathematics, the Mathematical Association of America, the National Council of Teachers of Mathematics, the American Mathematical Association of Two-Year Colleges, and The Institute of Management Sciences.

INTERMODULAR DESCRIPTION SHEET: UMAP Unit 572

TITLE: EVALUATING AND ANALYZING PROBABILISTIC FORECASTS

AUTHOR: J. Frank Yates
Department of Psychology
The University of Michigan
Ann Arbor, Michigan 48104

MATH FIELD: Probability and Statistics

APPLICATION FIELD: Forecasting

TARGET AUDIENCE: Students in a course on statistical analysis.

ABSTRACT: This module presents an analysis of the quality of probabilistic forecasts, such as weather forecasting and sporting events. It describes both discrete and continuous forecasts using graphical indicators. Reliability diagrams and covariance graphs are introduced. A variety of forecasting methods are evaluated.

PREREQUISITES: A first course in probability and statistics.

The UMAP Journal, Vol. V, No. 1, 1984

Evaluating and Analyzing Probabilistic Forecasts

J. Frank Yates
Department of Psychology
The University of Michigan
Ann Arbor, Michigan 48104

Table of Contents

1. INTRODUCTION . . . 1
  1.1 Some Practical Decision Problems . . . 1
  1.2 The Issue: Probabilistic Forecast Quality . . . 3
2. METHODS FOR DISCRETE FORECASTS . . . 4
  2.1 Situations and Notation . . . 4
  2.2 Graphical Indicators of Forecast Characteristics – Calibration-in-the-Small and Resolution . . . 8
  2.3 The Probability Score and Its Mean . . . 10
  2.4 The Sanders Decomposition of the Mean Probability Score . . . 11
  2.5 The Murphy Decomposition of the Mean Probability Score . . . 13
  2.6 Some Real Examples: Weather and Baseball . . . 15
3. METHODS FOR CONTINUOUS FORECASTS . . . 17
  3.1 Situations and Notation . . . 17
  3.2 Graphical Indicators of Forecast Characteristics – Calibration-in-the-Large, Slope, Scatter, Conditional Distributions . . . 17
  3.3 The Probability Score for Continuous Forecasts and the Covariance Decomposition of Its Mean . . . 22
  3.4 Weather and Baseball Revisited . . . 26
  3.5 Relationships Among the Decompositions of $\overline{PS}$. 28
4. PROPERNESS: A SPECIAL PROPERTY OF THE PROBABILITY SCORE . . . 28
5. BIBLIOGRAPHY . . . 31
6. EXERCISES . . . 32
7. ANSWERS TO EXERCISES . . . 37

Modules and Monographs in Undergraduate Mathematics and its Applications Project (UMAP)

The goal of UMAP was to develop, through a community of users and developers, a system of instructional modules in undergraduate mathematics and its applications that may be used to supplement existing courses and from which complete courses may eventually be built.

The Project was guided by a National Advisory Board of mathematicians, scientists, and educators. UMAP was funded by a grant from the National Science Foundation to the Consortium for Mathematics and Its Applications, Inc. (COMAP), a nonprofit corporation engaged in research and development in mathematics education.

The Project and author would like to thank Vera Lippitt of the University of Rochester and James Inglis of Bell Laboratories who reviewed this module and all others who assisted in the production of this unit.

1. Introduction

1.1 Some Practical Decision Problems

EXAMPLE 1: Consider the following lotteries:

Lottery 1:

Urn 1
800 Red
200 White
1000 Total

*Select a ball at random from Urn 1, which contains 800 red and 200 white balls.
*Win $5 if the selected ball is red; otherwise lose $1.

Lottery 2:

Urn 2
400 Red
600 White
1000 Total

*Select a ball at random from Urn 2, which contains 400 red and 600 white balls.
*Win $10 if the selected ball is red; otherwise lose $3.

Suppose a gambler is offered a choice of Lotteries 1 and 2. Which should she select?

The *expected value (EV)* or *expectation* of a lottery is the sum of all its possible prizes, each weighted by its probability. Thus, the EV of Lottery 1 is given by:

$$EV(L1) = P(\text{Red } 1)\ (\$5) + P(\text{White } 1)\ (-\$1)$$

$$= (.8)\ (\$5) + (.2)\ (-\$1)$$

$$= \$3.80.$$

Similarly, Lottery 2's EV can be shown to be $EV(L2) = \$2.20$. The *expectation (EV) maximization principle* says that one should always choose the option that has the higher EV. Thus, the principle would imply that Lottery 1 be chosen over Lottery 2, since $EV(L1) > EV(L2)$. Moreover, each should be chosen over doing nothing at all. This follows if one conceives of "doing nothing" as a degenerate gamble or "prospect" in which one "wins" \$0 no matter what event occurs, i.e., *EV*("Doing nothing") = \$0.

Why would one want to maximize *EV*? A number of arguments can be made both for and against the expectation maximization principle. A careful discussion of those arguments would carry us too far afield. Via the natural phenomena encapsulated in the "Law of Large Numbers," however, maximizing *EV* practically guarantees that in the long run one would be better off than one would be using any other decision rule. For instance, if the gambler were to play Lottery 1 100 times, her total earnings would almost certainly be somewhere close to $100 \times EV(L1) = \$380$. On the other hand, if she chose instead to play Lottery 2 100 times, her earnings would be in the vicinity of \$220.

EXAMPLE 2: The following *payoff matrix* summarizes the situation of a citrus grower:

Action	**Event**	
	$A =$ "$\leq 0°\ C$"	$\overline{A} =$ "$> 0°\ C$"
Protect Crop	$C = -\$1000$	$C = -\$1000$
Don't Protect Crop	$L = -\$8000$	\$0

The event (A) of interest to the grower is the occurrence of a fall in temperature to the freezing point. If the grower took precautionary protective measures, e.g., covering the plants, he would incur a cost of \$1000, regardless of whether event A or its complement $\overline{A}$ occurred. If no protective measures were taken and a freeze occurred, the grower would lose \$8000. If he took no action and no freeze occurred, he would lose nothing.

A *deterministic forecast* is a categorical assertion that a particular event will or will not occur. A *probabilistic forecast* is a numerical statement of the chance of an event's happening, with 1 indicating that the event definitely will occur, 0 indicating that it definitely will not, and numbers between 1 and 0 corresponding to intermediate degrees of certainty. The U.S. National Weather Service issues probabilistic forecasts of precipitation. In principle, they could do so for temperature, too. If the citrus grower in this problem had access to temperature probabilistic forecasts, he could apply the *EV* maximization principle to decide whether to protect or not protect his crop.

Let $p = P(A)$ be the probabilistic forecast of a freeze. Then the EV's of protecting and not protecting the crop would be,

$$EV(\text{Protect}) = pC + (1-p)C = C$$

and

$$EV(\text{Don't Protect}) = pL + (1-p)0 = pL,$$

where C is the cost of protection and L is the potential loss from a freeze when no protective action is taken. The grower should take protective action only if

$$EV(\text{Protect}) > EV(\text{Don't Protect}),$$

$$C > pL,$$

or

$$p > C/L = 1/8. \text{ (Remember that } L < 0.)$$

The probability $C/L = 1/8$ is called the *cutoff probability.*

EXAMPLE 3: You have already applied to several graduate schools and have the time and money to apply to only one more. You are considering Schools X and Y. You find School X more attractive than School Y. Your advisor tells you that you have a better chance of being admitted to School Y, however. To which school should you apply? If your advisor says that your chances of getting into School Y are only slightly better than your chances of getting into School X, you might be inclined to apply to School X anyway. However, you would probably do the opposite if your advisor indicated that your chances of being admitted to the two schools were substantially different.

1.2 The Issue: Probabilistic Forecast Quality

What is common to Examples 1 through 3 is that the decision maker trades off value against uncertainty. That is, good chances compensate for less attractiveness, and vice versa. As suggested by all the examples, it is useful to know not only whether an event is likely or unlikely to occur, but also *how* likely or unlikely its chances are. The application of systematic quantitative ways of making value-uncertainty tradeoffs, such as EV maximization, however, requires having quantitative measures of uncertainty, or probabilities.

A serious problem decision makers must confront is that not all probabilities are the same. In Example 1 there is little quarrel with setting $P(\text{Red } 1) = .8$ and $P(\text{Red } 2) = .4$. The probabilities of such "canonical" events are noncontroversial and are sometimes even called "objective." This is not so for the probabilities in Examples 2 and 3, or indeed for probabilistic forecasts in most real-world decision problems. In such situations there is seldom anything approaching an objective basis for generating probabilities. Thus, a forecaster must rely on historical and statistical information thought to be relevant to the event of interest. Or, he or she must use purely personal judgment.

Given that in many practical circumstances there is no standard way to arrive at probabilistic forecasts, it is not surprising when different forecasters offer different forecasts for the same event. The question which arises then is this: Whose forecasts should the decision maker use? For instance, in Example 3, suppose that, while your advisor says $P(\text{Admitted to } Y) > P(\text{Admitted to } X)$, another faculty member says $P(\text{Admitted to } X) > P(\text{Admitted to } Y)$. On whose judgment do you rely?

A decision maker would naturally prefer to use the probabilities provided by the best forecaster available. This requires that one have a way to assess the quality of probabilistic forecasts. One of the purposes of this module is to describe a technique for indexing overall probabilistic forecast accuracy. There are actually several different ways in which probabilistic forecasts can be good or poor. So, a second purpose of this module is to introduce procedures for analyzing differences in forecasting performance.

2. Methods for Discrete Forecasts

2.1 Situations and Notation

We will limit our attention to situations in which there is a single event A to be forecasted. It is assumed that the conditions for A's occurrence are replicable over many different occasions. For example, A might be "this student would be admitted to this graduate school" in a counseling situation; "rain will occur within 12 hours" in a weather forecasting situation; "this investment opportunity will be fruitful" in a financial decision situation; "this applicant would perform the job well" in an employment interviewing situation; and "this team will win" in a sports situation.

A total of N occasions on which event A may or may not occur are observed. On each of those occasions, the forecaster offers a probabilistic forecast $f(A)$ of A's occurrence, where $0 \leq f(A) \leq 1$. As with probabilities in general, $f(A) = 0$ indicates that the forecaster thinks A is certain *not* to occur; $f(A) = .5$ indicates that A and its complement $\bar{A}$ are judged equally likely to occur; and $f(A) = 1$ means that A is seen to be guaranteed to occur. Other values of $f(A)$ index intermediate degrees of certainty in A's occurrence.

In this section of the module, we restrict ourselves to circumstances under which the forecaster can offer only discrete forecasts. That is, on any one occasion, the forecaster can say only that the probability of A's occurrence is $f(A) = f_1, f_2, \ldots,$ or f_J. Alternatively, at the time the forecasts are made, the forecaster might be free to offer any forecast he or she pleases. After the fact, however, those forecasts might be rounded to the discrete forecasts $f_1, \ldots, f_J$.

We next define the function d, the *outcome index:*

$$
\begin{aligned}
d &= 1, \text{ if } A \text{ occurs} \\
&= 0, \text{ if } \bar{A} \text{ occurs, i.e., if } A \text{ does not occur.}
\end{aligned}
$$

It is convenient to index occasions according to the forecasts made for A on those occasions. Thus, occasion (i,j) is the ith occasion on which the forecaster offers forecast f_j. If the forecaster makes forecast f_j a total of N_j times, then it is clear that the forecaster effectively partitions the overall total of N occasions into J categories and that

$$N = \sum_{j=1}^{J} N_j.$$

The outcome index on the ith occasion when forecast f_j is given is denoted by d_{ij}.

Some concrete examples, even if hypothetical, should make clear how the conventions described above are used. These examples will also be used to illustrate the various concepts and measures to be presented below.

Imagine that our interest is in the abilities of several counselors to forecast whether students with particular credentials would be admitted to the graduate schools to which they apply. The counselors are presented with the same N cases, i.e., sets of student credentials, which are treated as forecasting occasions. To simplify things, we consider only a small number of cases, $N = 25$. The target event on each occasion is "student is admitted." Forecasts are restricted to tenths. So, for each student, the counselor must indicate that the probability of the student being admitted to the given school is .0, .1, ..., .9, or 1.0. After all admissions decisions are announced, the outcome indexes are known: $d = 1$ if a student is admitted; $d = 0$ otherwise. Table 1 presents the results of the exercise.

Table 1

Hypothetical Graduate School Admissions Forecasts and Outcomes

Student/ Occasion (k)	Outcome	Outcome Index (d_k)	Counselor Forecast (f_k) A	B	C	D
1	Admitted	1	.6	.5	.4	.8
2	Not Admitted	0	.4	.7	.7	.3
3	Admitted	1	.3	.6	.2	.7
4	Admitted	1	.4	.4	.4	.9
5	Not Admitted	0	.6	.5	.6	.8
6	Admitted	1	.7	.5	.9	.8
7	Not Admitted	0	.7	.7	.5	.1
8	Admitted	1	.7	.7	.3	.3
9	Not Admitted	0	.3	.4	.7	.7
10	Admitted	1	.5	.5	.6	.8
11	Not Admitted	0	.8	.6	.4	.3
12	Not Admitted	0	.6	.6	.5	.0
13	Admitted	1	.3	.4	.2	.8
14	Admitted	1	.6	.6	.3	.7
15	Not Admitted	0	.7	.5	.6	.3
16	Admitted	1	.1	.7	.3	.9
17	Admitted	1	.8	.6	.7	.7
18	Not Admitted	0	.3	.4	.6	.7
19	Not Admitted	0	.5	.6	.5	.3
20	Admitted	1	.8	.7	.4	.8
21	Not Admitted	0	.4	.5	.7	.2
22	Admitted	1	.4	.6	.2	.7
23	Admitted	1	.6	.7	.9	.3
24	Not Admitted	0	.1	.4	.6	.1
25	Admitted	1	.3	.6	.9	.7

Table 2 displays the same information as Table 1. However, the forecasts have been indexed according to the 11 possible forecast categories. In the section of the table assigned to each counselor, j denotes the pertinent category, where $j = 1$ corresponds to forecast $f_1 = .0$, $j = 2$ to forecast $f_2 = .1$, and so forth. The index i denotes the given individual forecasting occasion within a particular forecast category. For example, the fifth row in the Counselor A section of the table pertains to the third occasion ($i = 3$) on which the fourth potential forecast category ($f_4 = .3$) was offered. We see that the given student actually was admitted ($d_{34} = 1$). The fourth column in each section contains a statistic used below,

Table 2

Indexed Hypothetical Graduate School Admissions Forecasts and Outcomes

Counselor A				Counselor B				Counselor C				Counselor D			
i,j	f_j	d_{ij}	$\bar{d}_j$	i,j	f_j	d_{ij}	$\bar{d}_j$	i,j	f_j	d_{ij}	$\bar{d}_j$	i,j	f_j	d_{ij}	$\bar{d}_j$
1,2	.1	1	.50	1,5	.4	1	.40	1,3	.2	1	1.00	1,1	.0	0	.00
2,2	.1	0		2,5	.4	0		2,3	.2	1		1,2	.1	0	.00
1,4	.3	1	.60	3,5	.4	1		3,3	.2	1		2,2	.1	0	
2,4	.3	0		4,5	.4	0		1,4	.3	1	1.00	1,3	.2	0	.00
3,4	.3	1		5,5	.4	0		2,4	.3	1		1,4	.3	0	.33
4,4	.3	0		1,6	.5	1	.50	3,4	.3	1		2.4	.3	1	
5,4	.3	1		2,6	.5	0		1,5	.4	1	.75	3,4	.3	0	
1,5	.4	0	.50	3,6	.5	1		2,5	.4	1		4,4	.3	0	
2,5	.4	1		4,6	.5	1		3,5	.4	0		5,4	.3	0	
3,5	.4	0		5,6	.5	0		4,5	.4	1		6,4	.3	1	
4,5	.4	1		6,6	.5	0		1,6	.5	0	.00	1,8	.7	1	.71
1,6	.5	1	.50	1,7	.6	1	.62	2,6	.5	0		2,8	.7	0	
2,6	.5	0		2,7	.6	0		3,6	.5	0		3,8	.7	1	
1,7	.6	1	.60	3,7	.6	0		1,7	.6	0	.20	4,8	.7	1	
2,7	.6	0		4,7	.6	1		2,7	.6	1		5,8	.7	0	
3,7	.6	0		5,7	.6	1		3,7	.6	0		6,8	.7	1	
4,7	.6	1		6,7	.6	0		4,7	.6	0		7,8	.7	1	
5,7	.6	1		7,7	.6	1		5,7	.6	0		1,9	.8	1	.83
1,8	.7	1	.50	8,7	.6	1		1,8	.7	0	.25	2,9	.8	0	
2,8	.7	0		1,8	.7	0	.67	2,8	.7	0		3,9	.8	1	
3,8	.7	1		2,8	.7	0		3,8	.7	1		4,9	.8	1	
4,8	.7	0		3,8	.7	1		4,8	.7	0		5,9	.8	1	
1,9	.8	0	.67	4,8	.7	1		1,10	.9	1	1.00	6,9	.8	1	
2,9	.8	1		5,8	.7	1		2,10	.9	1		1,10	.9	1	1.00
3,9	.8	1		6,8	.7	1		3,10	.9	1		2,10	.9	1	

$$\bar{d}_j = (1/N_j) \sum_{i=1}^{N_j} d_{ij}, \tag{1}$$

the *mean outcome index* for the N_j occasions when forecast f_j was offered. Given the definition of d_{ij} as an indicator variable, it is clear that $\bar{d}_j$ is also the relative frequency with which students were admitted, i.e., event A occurred, when the forecast f_j had been given. For instance, 60% of the students who Counselor A indicated had a .3 probability of being admitted were *actually* admitted.

2.2 Graphical Indicators of Forecast Characteristics - Calibration-in-the-Small and Resolution

The extent to which probabilistic forecasts anticipate the events of interest overall is called their *external correspondence.* Roughly speaking, a collection of forecasts exhibits good external correspondence to the degree that high probabilities are assigned to events that ultimately occur, and low ones to those that do not.

External correspondence is not a unitary construct. A forecaster can fail to be externally correspondent in several meaningfully distinct ways. Two of the important aspects of external correspondence can be discerned from graphical displays called reliability diagrams. A reliability diagram is a plot of mean outcome indexes against the respective forecast categories. Figure 1 displays the reliability diagrams for the admissions forecasts of Counselors A through D. The number adjacent to each point is the number of forecasting occasions represented by the point.

The first dimension of external correspondence highlighted by reliability diagrams is calibration-in-the-small. A collection of forecasts is said to be *well-calibrated-in-the-small* to the extent that forecasts are matched by the corresponding relative frequencies with which the target event actually occurs, i.e., $f_j = \bar{d}_j$. If calibration-in-the-small were perfect, the target event would occur on 30% of the occasions on which forecast .3 was offered, on 65% of the occasions on which forecast .65 was offered, and so on. Graphically, calibration-in-the-small is indicated by the closeness of the points in the pertinent reliability diagram to the heavy 1:1 diagonal.

Figure 1 seems to indicate that Counselor B's forecasts were the best calibrated-in-the-small, although those of Counselor D were well-calibrated, too. Counselor A's

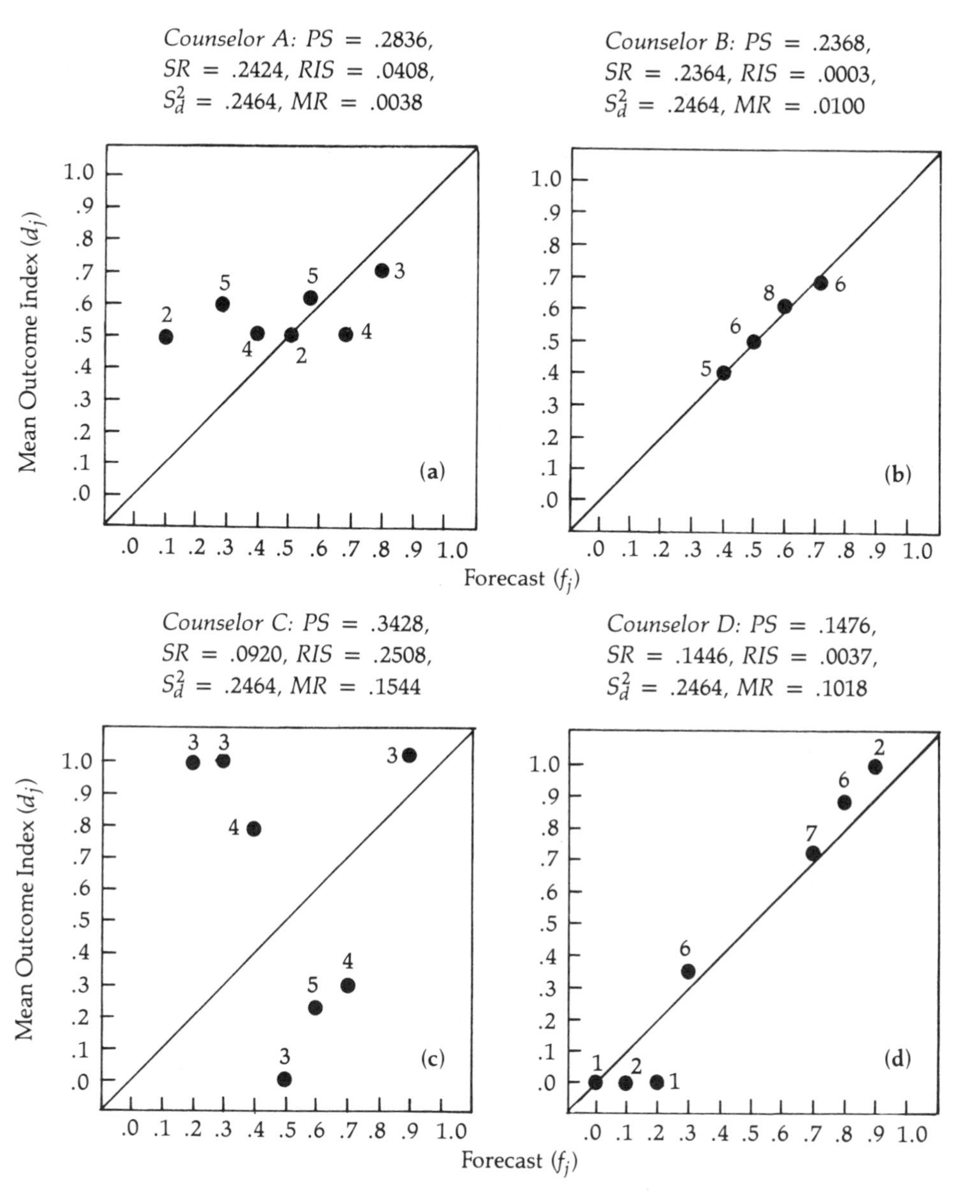

Figure 1. Reliability diagrams for the graduate admissions forecasts of Counselors A through D.

forecasts were poorly calibrated-in-the-small. The calibration-in-the-small of Counselor C's judgments was just awful.

The second aspect of external correspondence brought out by reliability diagrams is resolution. A collection of forecasts is *well-resolved* to the extent that occasions that result in the target event's occurrence and nonoccurrence are not assigned the same forecasts. Put another way, forecasts are perfectly resolved if the same judgment is never offered on two different occasions, one of which results in the target event's occurrence, the other of which does not. Good resolution is evidenced in a reliability diagram to the degree that the vertical coordinates of the points appear toward the top and bottom of the figure. Resolution is poorest when all the points have the same vertical coordinate, the overall proportion of the time the target event occurs.

Examining Figure 1, we see that the resolution of the forecasts of Counselors A and B was rather poor. Although Counselor C's calibration-in-the-small was terrible, his resolution was quite good. Counselor D's forecasts were both well-calibrated-in-the-small and fairly well-resolved. Intuitively, Counselor D seems to be the best admissions forecaster overall.

On first impressions, calibration-in-the-small is a most appealing quality for probabilistic forecasts. Perhaps one reason for this is that it seems to permit us to interpret the forecaster's probability statements the same way we might interpret the probabilities associated with such physical events as randomly drawing balls from urns. For instance, if one were repeatedly to sample (with replacement) from Urn 1 in Lottery 1 of Example 1, the relative frequency of red balls drawn should approach the assigned probability .8 in the long run.

Despite the apparent greater attractiveness of calibration-in-the-small, a case can be made for resolution being the more significant aspect of probabilistic forecasts, particularly those provided by human forecasters. Calibration concerns the forecaster's ability to assign *numbers* properly. In contrast, resolution pertains to the forecaster's ability to discriminate accurately among those occasions when the target event *will* occur from those when it *won't*. In principle, it seems that resolution reflects more fundamental insight than does calibration.

2.3 The Probability Score and Its Mean

It is useful to be able to summarize the external correspondence of probabilistic forecasts with a single number. Several measures for this purpose have been proposed and are employed in practice. The most common are the probability score and its mean.

Consider the ith occasion upon which the forecaster offers probability f_j for event A's occurrence. The *probability score* (PS) for that occasion is given by the formula

$$PS(f_j, d_{ij}) = (f_j - d_{ij})^2. \tag{2}$$

The forecaster's judgment is *perfect* if forecast 1.0 is offered when A occurs and forecast .0 when A does not occur; this results in PS = 0. The forecaster's judgment is *counterperfect* if forecast .0 is offered when A occurs and forecast 1.0 when it does not; this situation yields PS = 1. Intermediate values of PS indicate intermediate degrees of external correspondence.

Now consider the complete set of N forecasts. The *mean probability score* ($\overline{PS}$) for those N forecasts is defined by

$$\begin{aligned}\overline{PS}(f,d) &= (1/N) \sum_{j=1}^{J} \sum_{i=1}^{N_j} PS(f_j, d_{ij}) \qquad (3)\\ &= (1/N) \sum_{j=1}^{J} \sum_{i=1}^{N_j} (f_j - d_{ij})^2.\end{aligned}$$

The mean probability score has the same bounds and interpretations as the probability score. As an average, however, particularly over a large number of representative cases, it can be taken as a more reliable index of the forecaster's skills.

The mean probability scores of Counselors A through D are shown in the reliability diagrams in Figure 1. Low values of $\overline{PS}$ indicate good external correspondence. So, we see that, consistent with the impression created by the plots themselves, Counselor D is again indicated to have been the best forecaster, while Counselor C is measured as the worst.

2.4 The Sanders Decomposition of the Mean Probability Score

Frederick Sanders (1963), a meteorologist, has shown that $\overline{PS}$ can be expressed in the following way:

$$\overline{PS}(f,d) = (1/N) \sum_{j=1}^{J} N_j \bar{d}_j (1 - \bar{d}_j) + (1/N) \sum_{j=1}^{J} N_j (f_j - \bar{d}_j)^2. \tag{4}$$

It is instructive to consider how Equation (4) can be derived.

Recall that the summand in the formula for $\overline{PS}$ is $(f_j - d_{ij})^2$, which can be rewritten as $[(f_j - \bar{d}_j) + (\bar{d}_j - d_{ij})]^2$. Squaring the binomial and taking the mean, we have

$$\overline{PS}(f,d) = (1/N) \sum_{j=1}^{J} \sum_{i=1}^{N.} (f_j - \bar{d}_j)^2 \tag{5}$$

$$+ 2(1/N) \sum_{j=1}^{J} \sum_{i=1}^{N_j} (f_j - \bar{d}_j)(\bar{d}_j - d_{ij})$$

$$+ (1/N) \sum_{j=1}^{J} \sum_{i=1}^{N_j} (\bar{d}_j - d_{ij})^2$$

$$= A + B + C.$$

As far as the second summation in term A is concerned, $(f_j - \bar{d}_j)^2$ is a constant. So, it is clear that

$$A = (1/N) \sum_{j=1}^{J} N_j (f_j - \bar{d}_j)^2.$$

It is left as an exercise to show that $B = 0$. Term C can be written as

$$C = (1/N) \sum_{j=1}^{J} N_j (1/N_j) \sum_{i=1}^{N_j} (\bar{d}_j - d_{ij})^2.$$

But

$$(1/N_j) \sum_{i=1}^{N_j} (\bar{d}_j - d_{ij})^2$$

is the variance of d_{ij}, $S^2_{d_{ij}}$.

It is left as another exercise to demonstrate that

$$S^2_{d_{ij}} = \bar{d}_j(1 - \bar{d}_j).$$

Thus

$$C = (1/N) \sum_{j=1}^{J} N_j \bar{d}_j (1 - \bar{d}_j),$$

and the result has been shown.

Equation (4) is called the *Sanders decomposition of* $\overline{PS}$. The first term in the decomposition is the *Sanders resolution* for the collection of N forecasts. As its name and form suggest, it indexes resolution. Again recognizing that the forecaster's aim is to minimize $\overline{PS}$, the Sanders resolution should be minimized, too.

The Sanders resolution will achieve its minimum of zero when each summand $N_j\bar{d}_j(1-\bar{d}_j)$ is zero. This occurs (a) when $N_j = 0$, i.e., when forecast f_j is never given; (b) when $\bar{d}_j = 0$, i.e., when event A never occurs when forecast f_j is offered; or (c) when $\bar{d}_j = 1$, i.e., when event A *always* occurs when forecast f_j is offered. That is, as indicated before, resolution is perfect when the forecaster never gives the same forecast on two different occasions, one of which results in event A's occurrence, the other of which does not.

The Sanders resolution *(SR)* scores for Counselors A through D are shown in their reliability diagrams in Figure 1. The scores probably conform to your subjective impression of the relative resolution of the four sets of forecasts: Counselor C exhibited the best resolution; Counselor A and Counselor B manifested rather poor resolution; Counselor D's resolution was more like that of Counselor C than that of Counselors A and B.

The second term in the Sanders decomposition is called the *reliability-in-the-small.* The reliability-in-the-small clearly indexes calibration-in-the-small. It is zero when calibration is perfect, since then $f_j = \bar{d}_j$, for all j.

The reliability-in-the-small scores for the admissions forecasts of Counselors A through D are also shown in Figure 1, labeled *RIS*. Once more, the scores are consistent with the impressions conveyed by the plots: The reliability of Counselor B's forecasts is best, followed in order by that of Counselors D, A, and C. Except for rounding error, the scores shown in the graphs agree, as they must, with the prescription of the Sanders decomposition: $\overline{PS} = SR + RIS$. It is of some interest to note that almost all of Counselor B's $\overline{PS}$ is due to poor resolution rather than poor calibration, whereas the reverse is true for Counselor C. The best overall forecaster, Counselor D, exhibits more balance.

2.5 The Murphy Decomposition of the Mean Probability Score

Allan Murphy (1973), another meteorologist, has demonstrated that the Sanders resolution can be represented in an especially useful fashion:

$$(1/N) \sum_{j=1}^{J} N_j\bar{d}_j(1-\bar{d}_j) = \bar{d}(1-\bar{d}) - (1/N) \sum_{j=1}^{J} N_j(\bar{d}_j-\bar{d})^2. \quad (6)$$

In Equation (6), $\bar{d}$ is the *overall mean outcome index* and is defined by

$$\bar{d} = (1/N) \sum_{j=1}^{J} \sum_{i=1}^{N_j} d_{ij}$$

Also known as the *base rate*, $\bar{d}$ is the overall relative frequency of the target event's occurrence. The derivation of Equation (6) requires a special insight, but is then rather direct. It is left as one of the exercises at the end of the module (along with a hint).

The usefulness of Equation (6) arises from the fact that $\bar{d}(1-\bar{d})$ is the variance S_d^2 of the outcome index d. Thus, it is controlled by whatever determines event A's occurrence or nonoccurrence; it is *not* affected by the forecaster's judgments. Accordingly, the second term on the right hand side of Equation (6), called the *Murphy resolution*, is that part of the Sanders resolution that *is* under the forecaster's control. Bearing in mind that the forecaster cannot influence $\bar{d}(1-\bar{d})$, we realize that the Sanders and Murphy resolutions convey the same information about the forecaster's powers to discriminate occasions when event A will and will not occur. Notice that the Murphy resolution enters Equation (6) negatively. Therefore, the Murphy resolution is maximized when the Sanders resolution is minimized, and vice versa, given the same forecasting situations.

The significance of the separation of the forecaster-controlled and -uncontrolled parts of the Sanders resolution in Equation (6) is important if one wishes to compare resolving powers when forecasts are made on different collections of forecasting situations for which the target events' relative frequencies are not the same. Consider an example: Weather Forecaster 1 offers a series of precipitation forecasts in an area in the Pacific Northwest. His Sanders resolution is $SR_1 = .1515$. Weather Forecaster 2 offers similar forecasts in a Midwestern area, achieving a Sanders resolution of $SR_2 = .2134$. On the basis of the Sanders resolutions, it appears that Forecaster 1 has better discrimination ability. But now take into account the relative frequencies of precipitation, $\bar{d}_1 = .8$ in Forecaster 1's area, $\bar{d}_2 = .4$ in Forecaster 2's. Thus, the respective outcome index variances are

$$S_{d_1}^2 = .1600 \text{ and } S_{d_2}^2 = .2400.$$

If the Murphy resolution is abbreviated by MR, we see that $MR_1 = .1600 - .1515 = .0085$, whereas $MR_2 = .2400 - .2134 = .0266$. That is, Forecaster 2 has better, not worse, discrimination ability.

If Equation (6) is substituted into Equation (4), one obtains the *Murphy decomposition of* $\overline{PS}$:

$$\overline{PS}(f,d) = \bar{d}(1-\bar{d}) - (1/N) \sum_{j=1}^{J} N_j(\bar{d}_j - \bar{d})^2 + (1/N) \sum_{j=1}^{J} N_j(f_j - \bar{d}_j)^2. \qquad (7)$$

Figure 1 contains scores for all the components of the Murphy decomposition of $\overline{PS}$ for Counselors A through D. Because all the counselors considered the same cases, nothing new about the forecasters' abilities is conveyed by these components as compared to the components in the Sanders decomposition.

2.6 Some Real Examples: Weather and Baseball

The procedures and measures described above have been applied to probabilistic forecasting in several real-world contexts. To permit an appreciation for the sorts of forecasting performance one might observe under such conditions, the results of two actual forecasting analyses are presented here. Both analyses are from a paper by Yates (1981).

The first analysis was applied to weather forecasting data reported by Murphy and Winkler (1977). Murphy and Winkler's Forecaster A made a large number of probabilistic forecasts in the Chicago area. The target event was "precipitation will occur" within a certain period of time. Figure 2(a) presents the reliability diagram and performance statistics for Forecaster A's precipitation forecasts.

The second analysis was applied to baseball forecasts made by the well-known professional oddsmaker, James (Jimmy-the-Greek) Snyder. Each day during the baseball season, Jimmy-the-Greek publishes in newspapers around the country his odds on the games being played that day. These odds can be converted directly into probabilistic forecasts. For purposes of analysis, the target event A was designated as "home team wins." For example, suppose Jimmy-the-Greek reported that Detroit was favored 2:1 over Kansas City at home. This means that $P(A)/[1-P(A)] = 2$, where $P(A)$ is Jimmy-the-Greek's probabilistic forecast that the home team, Detroit, would defeat the visitor, Kansas City. Thus, $P(A) = 2/3$. Forecasts for a total of 562 consecutive games were considered.

Figure 2(b) shows the reliability diagram and performance statistics for Jimmy-the-Greek's baseball forecasts.

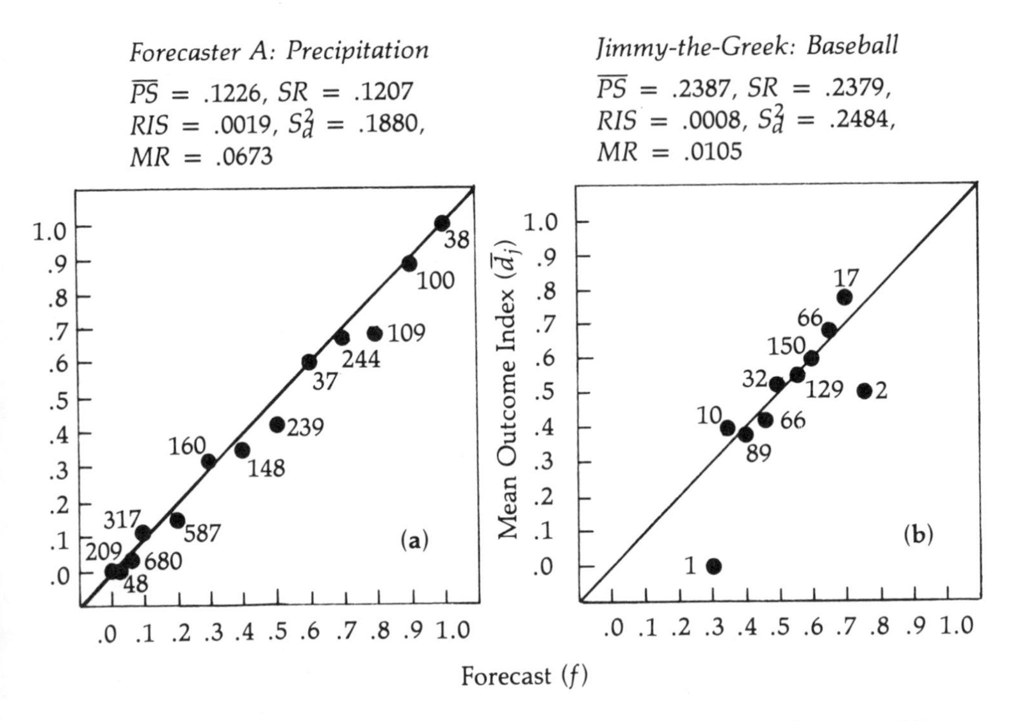

Figure 2. Reliability diagrams for Forecaster A's precipitation forecasts and Jimmy-the-Greek's baseball forecasts.

Judging from the reliability diagrams as well as the reliability-in-the-small scores, both Forecaster A's and Jimmy-the-Greek's forecasts were well-calibrated-in-the-small. The reliability diagrams suggest that Forecaster A's calibration was better, whereas the reliability scores indicate the converse. The reliability diagrams are a little deceptive, however. The impression of Jimmy-the-Greek's miscalibration is unduly influenced by the two outlier points, which are based on a total of only three cases. The reliability-in-the-small measure takes proper account of this low frequency.

The reliability diagrams also indicate that Jimmy-the-Greek was a much more conservative prognosticator than was Forecaster A. Whereas Forecaster A used the entire probability scale, Jimmy-the-Greek never offered forecasts outside the 30-75% range. The graphs appear to indicate that Jimmy-the-Greek's forecasts were much less well-resolved than Forecaster A's. The resolution difference is indicated whether one considers either the Sanders or the Murphy resolution scores.

Overall, Forecaster A was much better at anticipating precipitation than was Jimmy-the-Greek at predicting baseball game winners, as confirmed by the $\overline{PS}$ values. However, from the point of view of relative frequencies of the target event's occurrence ($\bar{d}$ = .251 for precipitation and $\bar{d}$ = .539 for home-team baseball wins), Jimmy-the-Greek's task was more difficult than Forecaster A's. There is more uncertainty in an event with a relative frequency closer to .5 rather than to either extreme of .0 or 1.0. Nevertheless, even taking this into account via the Murphy decomposition, we see that the forecaster-controlled portions of $\overline{PS}$ favor Forecaster A.

3. Methods for Continuous Forecasts

3.1 Situations and Notation

As far as substantive content is concerned, the forecasting situations of interest now are the same as those considered previously. The only difference is that, with so-called *continuous* rather than discrete forecasts, any number of distinct forecast values between 0 and 1 are treated. The forecaster is not required to restrict his or her forecasts to any specific possibilities.

The N forecasting occasions under consideration are indexed by $k = 1, \ldots, N$. The forecast offered on occasion k is denoted by f_k, the outcome index by d_k. Table 1 provides an illustration of such indexing. The same data will be used to illustrate continuous procedures as were used to demonstrate discrete ones. Now, we simply ignore the fact that the forecasts happen to fall into a limited set of categories.

3.2 Graphical Indicators of Forecast Characteristics — Calibration-in-the-Large, Slope, Scatter, Conditional Distributions

As implied previously, one *could* analyze continuous forecasts by rounding them into discrete forecasts and applying the methods described above. An alternative strategy is described presently.

The graphical phase of the approach requires the construction of *covariance graphs*. A covariance graph is similar to an ordinary scatterplot. Outcome index values are displayed on the abscissa, forecasts on the ordinate. Since the outcome index can assume only two distinct values, there will generally be many multiple points. These are represented as horizontal bars of lengths corresponding to frequencies. Thus, effectively, two *conditional forecast distribution* histograms are constructed, one for forecasts of event *A* conditional on event *A*'s occurrence, the other for such forecasts conditional on event *A*'s nonoccurrence. Figure 3 displays the covariance graphs for the graduate admissions forecasts of Counselors A through D. The number in parentheses on the longest bar in each histogram is the number of occasions represented by that bar.

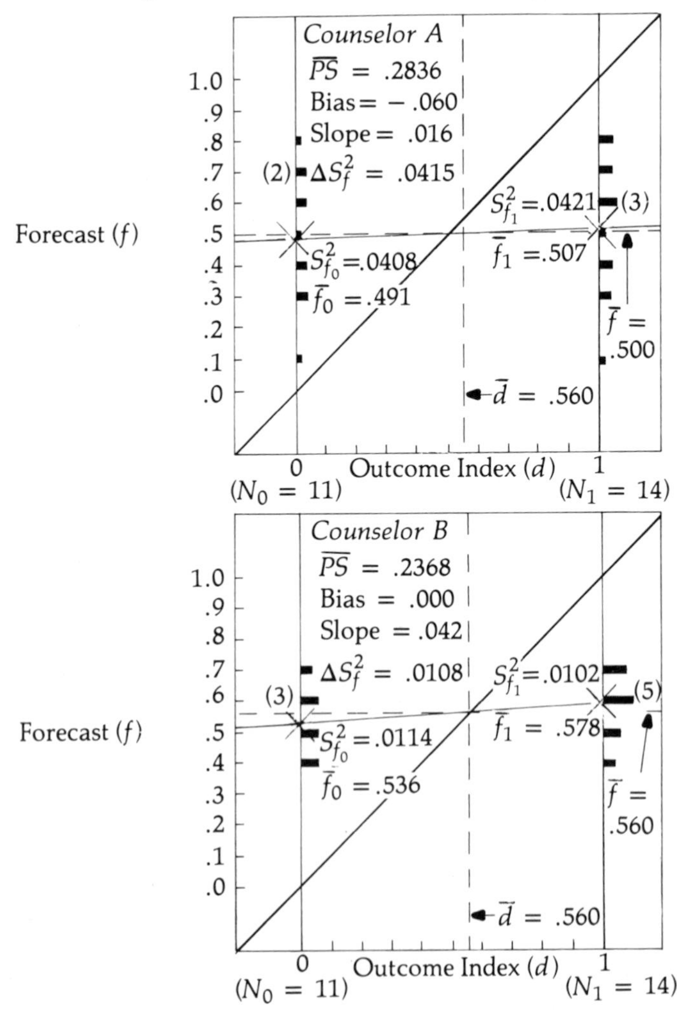

Figure 3. The covariance graphs for graduate admissions forecasts of Counselors A through D (continued on next page).

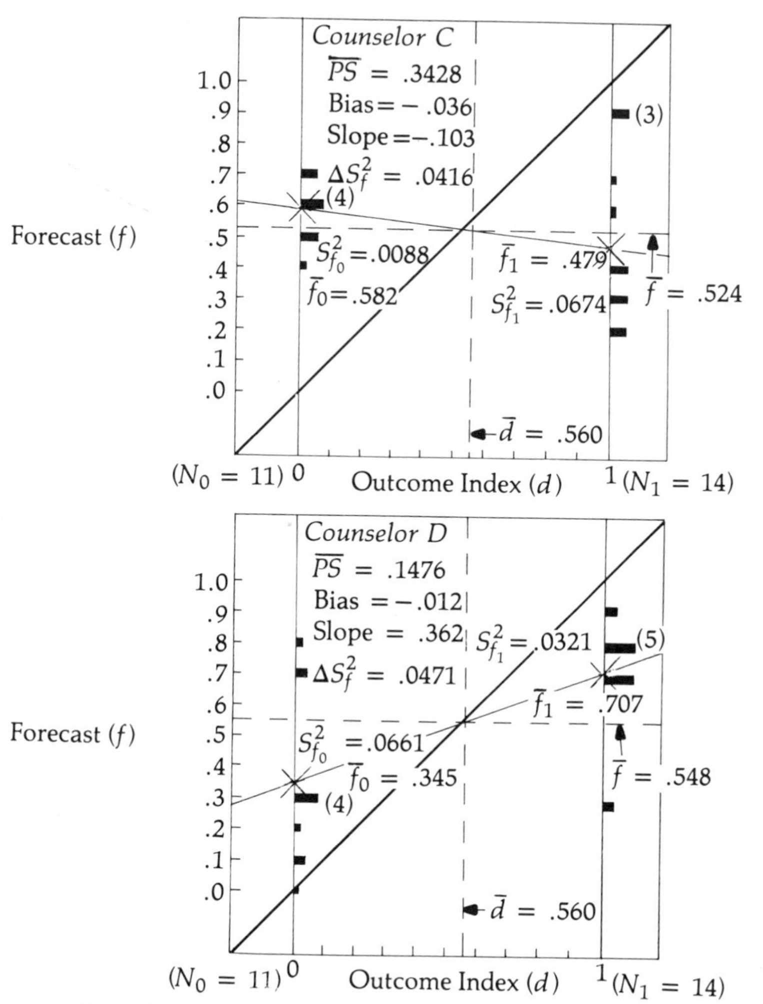

Figure 3 (continued). The covariance graphs for graduate admissions forecasts of Counselors A through D.

Covariance graphs provide a different perspective on the information contained in reliability diagrams. For certain kinds of insight, reliability diagrams seem to be more useful; for others, covariance graphs. As will become apparent, covariance graphs generally entail a somewhat more fine-grained level of analysis, however.

In each graph in Figure 3 there are vertical and horizontal dotted lines. The vertical line passes through the point $(\bar{d},0)$, the horizontal through the point $(0,\bar{f})$, where

$$\bar{d} = (1/N) \sum_{k=1}^{N} d_k \text{ and } \bar{f} = (1/N) \sum_{k=1}^{N} f_k.$$

Forecasts are said to be *well-calibrated-in-the-large* to the extent that $\bar{f}$ approximates $\bar{d}$, that is, the extent to which the mean overall forecast matches event *A*'s relative frequency of occurrence.

The *bias* of the forecasts is defined as $\bar{f} - \bar{d}$. If a collection of forecasts were unbiased or perfectly calibrated-in-the-large, this fact would be evident in the covariance graph by the intersection of the "$\bar{d}$ line" and the "$\bar{f}$ line" appearing on the heavy 1:1 diagonal, whose equation is $f = d$. The magnitude of a given bias is indicated by the vertical distance of the intersection point $(\bar{d},\bar{f})$ from the 1:1 diagonal. The bias is positive if $(\bar{d},\bar{f})$ is above the diagonal and negative if it is below.

We see from the lines and bias measures in Figure 3 that Counselor B was the only one whose forecasts were perfectly calibrated-in-the-large. To varying degrees, the forecasts of each of the other counselors were negatively biased. That is, they underestimated students' chances of being admitted to the graduate schools to which they applied. Calibration-in-the-large can sometimes be detected in reliability diagrams when the points are generally displaced to the left or right of the diagonal. A reexamination of Figure 1 suggests, however, that miscalibration-in-the-large is ordinarily more easily recognized in covariance graphs.

The second important forecasting performance dimension that can be readily seen in covariance graphs is *covariation* between forecasts and outcome indexes. Consistent with the very concept of external correspondence, the forecaster should strive to have high probabilities associated with $d = 1$, and low ones with $d = 0$. As will be shown below, the essential component of the statistical covariance of forecasts and outcome indexes is the difference between the *conditional mean forecast* for event *A* when *A* actually occurs (denoted $\bar{f}_1$) and that when *A* does not occur (denoted $\bar{f}_0$). The points $(0,\bar{f}_0)$ and $(1,\bar{f}_1)$ are designated by *X*'s in covariance graphs, as exemplified in Figure 3. The line connecting these points is technically the regression line of forecasts on outcome indexes. The number $\bar{f}_1 - \bar{f}_0$ is the *slope* of that regression line and can be taken as an indicator of the overall covariation of forecasts and outcome indexes. With the exception of Counselor D, the forecast slopes for the admissions counselors were rather modest. That of Counselor C was perversely negative.

How might one interpret forecast slopes? In some sense, the slope seems to get at the core of the forecaster's judgment ability. Perhaps it reflects the forecaster's access, sensitivity, and proper interpretation of signs that have some diagnostic value for event *A*'s occurrence. Otherwise how could the forecaster reliably evidence a nontrivial slope, particularly one that is properly oriented?

The third important forecasting performance characteristic brought out by covariance graphs is the *scatter* of forecasts about the conditional means. One possible reason forecasts might be widely dispersed around the conditional means is that the forecaster improperly responds to indicators *believed* to be diagnostic of event *A*'s occurrence, but which really are *not* diagnostic. Another potential reason is that even the best indicators provided by the forecasting situation itself simply are not very diagnostic. Yet another factor that might contribute to the wide scattering of forecasts is the forecaster's incentive structure. If the forecaster is heavily penalized for extreme errors, he or she is unlikely to make extreme forecasts.

One can gain a general impression of the scatter of the forecasts of Counselors A through D from Figure 3. Each of the graphs also contains a measure of that scatter, $\triangle S_f^2$, to be defined explicitly below. The forecasts of Counselors A, C, and D were all more or less equivalent with respect to scatter. Counselor B's forecasts were considerably less dispersed.

A final class of forecasting performance features evident from covariance graphs are conditional forecast *distribution differences*. The most obvious and important of such potential differences pertains to location, as indexed by the slope. There are sometimes other marked differences, too. One such difference is in dispersion. The variability of forecasts when event *A* occurs may differ greatly from that when *A* does not occur. Extending reasoning described above, this might mean that nature provides more "foils" as supposed accurate indicators for event *A* than for event $\overline{A}$, or vice versa. Another difference that might be of interest is shape. It sometimes happens that one conditional forecast distribution, for example, is skewed, while the other is symmetric.

The essential point to bear in mind is that *any* reliable distinction between what the forecaster does when event A occurs as compared to when it does not is an indication of the forecaster's ability to anticipate the future. Such ability can then be studied and possibly put to use. A careful examination of covariance graphs can reveal such distinctions. Consider again Counselor C's covariance graph for admissions forecasts. Counselor C was the one with excellent resolution. That individual's resolution manifests itself in the covariance graph by the almost complementary character of the conditional forecast distributions; forecasts made when admission was granted were seldom made when admission was denied, and conversely. Thus, whether consciously or not, Counselor C somehow "knew" almost perfectly when a student would or would not be admitted. He was just not very good—indeed, was perversely incompetent—at translating that knowledge into expressed forecasts that have the traditional interpretation. But that seems to be easily "learnable." Moreover, if Counselor C's performance pattern were reliable, a person finding that Counselor C assigned a probability of .3 to admission could be practically assured that admission would actually be granted, and conversely when a probability of .6 was assigned.

3.3 The Probability Score for Continuous Forecasts and the Covariance Decomposition of Its Mean

The continuous forecast version of the *probability score* for occasion k is given by

$$PS(f_k, d_k) = (f_k - d_k)^2. \tag{8}$$

The *mean probability score* is then defined as

$$\begin{aligned} \overline{PS}(f,d) &= (1/N) \sum_{k=1}^{N} PS(f_k, d_k) \\ &= (1/N) \sum_{k=1}^{N} (f_k - d_k)^2. \end{aligned} \tag{9}$$

The interpretations of PS and $\overline{PS}$ are the same as before.

The following *covariance decomposition of* $\overline{PS}$ can also be derived:

$$\overline{PS}(f,d) = \bar{d}(1-\bar{d}) + S^2_{f,\,\min} + \triangle S^2_f + (\bar{f}-\bar{d})^2 - 2S_{fd}, \tag{10}$$

where $\bar{d}$ and $\bar{f}$ are, as defined previously, the overall mean outcome index and mean forecast; S_{fd} is the covariance of forecasts and outcome indexes; and $S^2_{f,\,\min}$ and $\triangle S^2_f$ form a partition of the variance of the forecasts, S^2_f, to be described in more detail below. Equation (10) is a slight reworking of a well-known way of expressing a mean squared error in terms of means, variances, and the covariance:

$$(1/N) \sum_{k=1}^{N} (f_k - d_k)^2 = S^2_d + S^2_f + (\bar{f}-\bar{d})^2 - 2S_{fd}. \tag{11}$$

The demonstration of Equation (11) is left as an exercise. Let us now consider each of the terms on the right-hand side of Equation (10).

As indicated previously, and as apparent from a comparison of Equations (10) and (11), the quantity $\bar{d}(1-\bar{d})$ is the variance of the outcome index and is out of the forecaster's control. The fourth term on the right-hand side of Equation (10), $(\bar{f}-\bar{d})^2$, the square of the bias, is called the *reliability-in-the-large.* The reliability-in-the-large indexes calibration-in-the-large the same way that the reliability-in-the-small indexes calibration-in-the-small.

Imagine a constant forecaster who always offers the same forecast, $f_k = c$. Equations (10) and (11) make it clear that, for this forecaster,

$$\overline{PS}(f,d) = \overline{PS}(c,d) = \bar{d}(1-\bar{d}) + (c-\bar{d})^2. \tag{12}$$

It is apparent from Equation (12) that no constant forecaster can do better than the "relative frequency forecaster" who always offers $f_k = \bar{d}$, and thus earns the score $\overline{PS}(\bar{d},d) = \bar{d}(1-\bar{d})$. It requires a special kind of baseline knowledge to be able to anticipate what $\bar{d}$ will be. For example, in the graduate admissions illustration, a counselor would have to have considerable experience to know that the base rate for graduate admissions with the sorts of students under consideration is in the vicinity of 56%. So, $\bar{d}(1-\bar{d})$ provides a very useful point of reference for interpreting $\overline{PS}$ values.

Another point of reference is offered by the performance of the "uniform forecaster." The uniform forecaster behaves as if all the potential events were equally likely. In the situations considered here, this means that he or she always offers the constant forecast $f_k = .5$. Equation (12) then indicates that the uniform forecaster would achieve the score .25.

For the graduate admissions forecasts, $\bar{d} = .56$, yielding $\bar{d}(1-\bar{d}) = .2464$. Figures 1 and 3 reveal that, while Counselors B and D earned $\overline{PS}$ scores superior to (lower than) those of either the relative frequency or the uniform forecaster, Counselors A and C did not. Thus, in terms of $\overline{PS}$, one could do better by paying no attention whatsoever to the forecasts of Counselors A and C and simply using even odds. At first blush, this is surprising. However, an examination of Equation (10) makes clear the particular ways a forecaster can indeed do worse than to ignore distinctions between forecasting occasions and offer constant forecasts. One should also bear in mind that, despite the present remarks, Counselor C's forecasts *do* indicate significant anticipatory insight.

The *covariance* of f and d is defined by

$$S_{fd} = (1/N) \sum_{k=1}^{N} (f_k-\bar{f})(d_k-\bar{d}).$$

The covariance is related to the *product-moment correlation coefficient* r_{fd} by

$$r_{fd} = \frac{S_{fd}}{S_f S_d} .$$

Recall that d is a dichotomous indicator variable. Thus, r_{fd} in this case is the *point-biserial correlation coefficient.* It is left as an exercise to show that the covariance can therefore be written as

$$S_{fd} = (\bar{f}_1 - \bar{f}_0)\bar{d}(1-\bar{d}), \quad (13)$$

where the forecast means conditional upon event A's occurrence and nonoccurrence are, respectively,

$$\bar{f}_1 = (1/N_1) \sum_{m=1}^{N_1} f_{1m}$$

and

$$\bar{f}_0 = (1/N_0) \sum_{n=1}^{N_0} f_{0n}.$$

In the preceding definitions, f_{1m} represents forecasts associated with the N_1 occasions when event A actually occurs, while f_{0n} represents forecasts associated with the N_0 occasions when event A does not occur, $N = N_0 + N_1$. As it had to be, the covariance is thus directly related to the slope of the forecast-outcome index regression line. Also consistent with our intuitions and Equation (10), the larger the slope, the better is the forecasting performance.

In terms of improving $\overline{PS}$, Equation (11) suggests that the forecaster should try to minimize the overall forecast variance, S_f^2. There is an obvious qualification on the advisability of pursuing this goal, however. The variance S_f^2 could be reduced to zero by making constant forecasts. This eliminates S_{fd} too, however. So, a conditional goal might be substituted: Take the covariance as a *given* indicator of the forecaster's ability to discriminate individual forecasting situations usefully. Then seek to minimize S_f^2 while maintaining that covariance. Equation (10) suggests a way to think about and evaluate success at this strategy.

It is apparent from Equations (10) and (11) that

$$S_f^2 = S_{f,min}^2 + \triangle S_f^2,$$

where $S_{f,\mathrm{min}}^2$ is defined by

$$S_{f,\mathrm{min}}^2 = (\bar{f}_1 - \bar{f}_0)^2 \bar{d}(1-\bar{d}). \quad (14)$$

It is left as an exercise to show that $S^2_{f,\min}$, the *conditional minimum forecast variance*, is the smallest value of S^2_f one can achieve, given particular values of $\bar{f}_1$, $\bar{f}_0$, and $\bar{d}$. Graphically,

$$S^2_f = S^2_{f,\min}$$

when the covariance graph reveals that the forecaster always indicates $f_k = \bar{f}_1$ when event A occurs and always reports $f_k = \bar{f}_0$ when event A does not occur. In other words, there is no scatter around the conditional forecast means. The quantity $\triangle S^2_f$ is whatever "excess" variance exists in the forecasts beyond that necessary to achieve the covariance. It is left as another exercise to show that $\triangle S^2_f$ is a weighted mean of the conditional forecast variances:

$$\triangle S^2_f = \frac{N_0 S^2_{f0} + N_1 S^2_{f1}}{N} \tag{15}$$

where

$$S^2_{f_0} = (1/N_0) \sum_{n=1}^{N_0} (f_{0n} - \bar{f}_0)^2$$

and

$$S^2_{f_1} = (1/N_1) \sum_{m=1}^{N_1} (f_{1m} - \bar{f}_1)^2.$$

It should now be clear how and why a forecaster might do worse than providing constant forecasts. As suggested by Equation (10), the forecaster might be sensitive to reliably diagnostic signs of event A's occurrence and nonoccurrence, but interpret them in the wrong way, thus manifesting a negative covariance. Alternatively, the forecaster might introduce excessive scatter into his or her forecasts, for any of the reasons suggested previously, e.g., responsiveness to nondiagnostic signs, weak incentives, or even simple random error.

For the sake of completeness, Table 3 presents the invert covariance decompositions of $\overline{PS}$ for the graduate admissions forecasts of Counselors A through D.

3.4 Weather and Baseball Revisited

Figure 4 displays the covariance graphs and associated statistics for Forecaster A's precipitation forecasts and Jimmy-the-Greek's baseball forecasts. The covariance graphs seem to reveal more dramatic differences in forecasting performance than were apparent from the corresponding reliability diagrams.

Jimmy-the-Greek's calibration was almost astonishing— $\bar{f}$ = 53.8% vs. $\bar{d}$ = 53.9%. In contrast, Forecaster A exhibited a clear bias of 3% toward overpredicting the occurrence of precipitation. It seems plausible that this bias might be a reflection of the perceived relative costs of over- and underpredicting precipitation. Failing to carry an umbrella when it rains is more disagreeable than carrying an umbrella when it does not. One would probably react accordingly to weather forecasters who lead us to commit such errors.

The slope difference between Forecaster A and Jimmy-the-Greek is quite remarkable, particularly since the difference cannot be detected in the reliability diagrams. Exactly *why* there is such a difference is an interesting question. The difference might mean that Forecaster A is simply a better judge of the future than Jimmy-the-Greek.

Table 3

Covariance Decompositions of $\overline{PS}$ for Hypothetical and Real Illustrations

Forecaster	Decomposition Components										
	$\overline{PS}$	=	$\bar{d}(1-\bar{d})$	+	$S^2_{f,\min}$	+	ΔS^2_f	+	$(\bar{f}-\bar{d})^2$	− 2	× S_{fd}
Hypothetical											
Counselor A	.2836	=	.2464	+	.0001	+	.0415	+	.0036	− 2 ×	(.0039)
Counselor B	.2368	=	.2464	+	.0004	+	.0108	+	0	− 2 ×	(.0104)
Counselor C	.3428	=	.2464	+	.0026	+	.0416	+	.0013	− 2 ×	(−.0254)
Counselor D	.1476	=	.2464	+	.0322	+	.0471	+	.0001	− 2 ×	(.0891)
Real											
Forecaster A	.1223	=	.1880	+	.0264	+	.0483	+	.0009	− 2 ×	(.0706)
Jimmy-the-Greek	.2383	=	.2484	+	.0003	+	.0078	+	0^+	− 2 ×	(.0092)

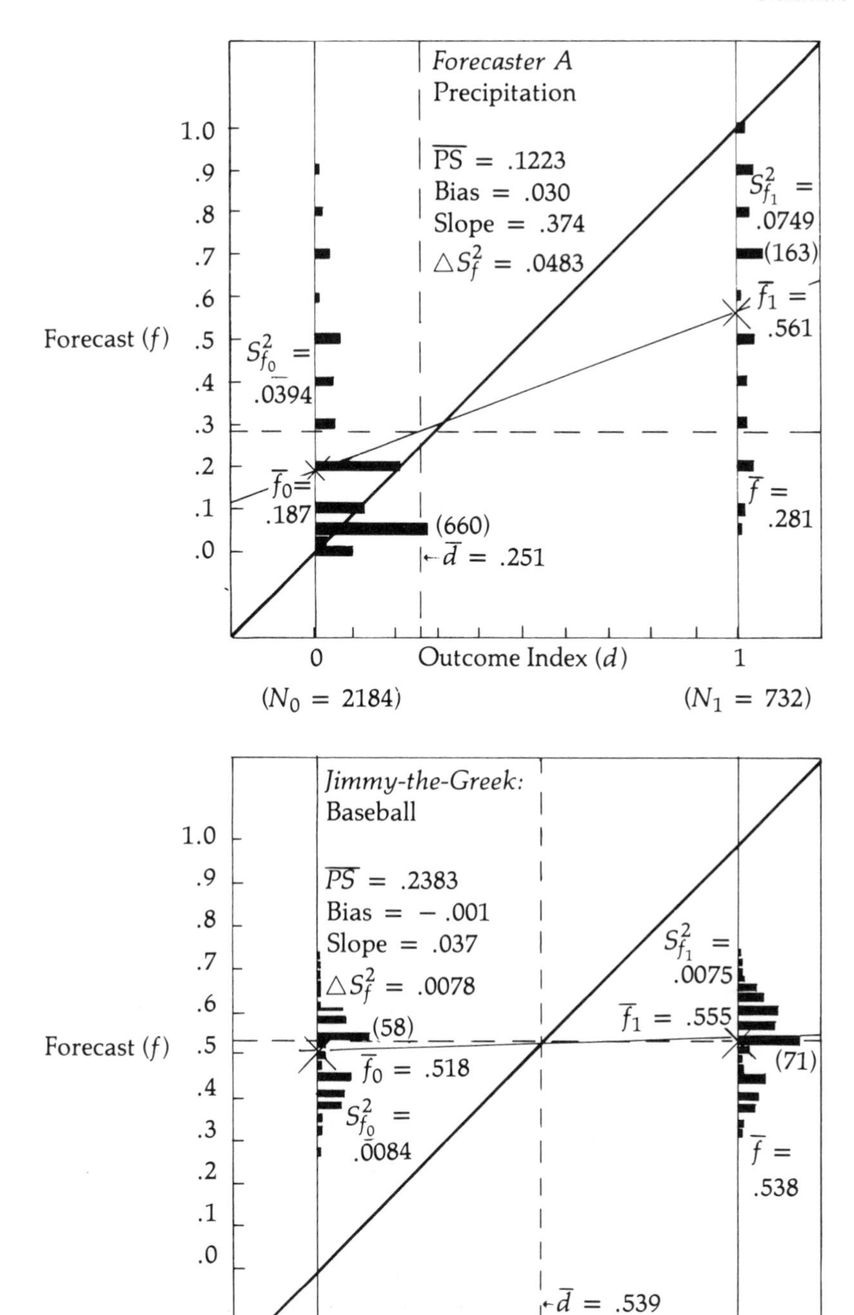

Figure 4. Covariance graphs for Forecaster A's precipitation forecasts and Jimmy-the-Greek's baseball forecasts.

It might be a reflection of their different incentive structures, too. Undoubtedly, the difference is also mediated by the fact that one context is weather forecasting, the other sports forecasting. Sports forecasting might be inherently more difficult; diagnostic signs might be less available to *anyone*.

There are also distinct differences in the conditional forecast distributions of Forecaster A and Jimmy-the-Greek. Jimmy-the-Greek's distributions are much tighter, perhaps as a reflection of incentives. Also note that Jimmy-the-Greek's distributions are both fairly symmetric, whereas one of Forecaster A's distributions is essentially symmetric, the other skewed. One cannot help being impressed by the wide dispersion of Forecaster A's precipitation forecasts when precipitation actually occurred. It would be of some interest to know why this happened.

Again for completeness, Table 3 summarizes the covariance decomposition statistics for Forecaster A's and Jimmy-the-Greek's forecasts.

3.5 Relationships Among the Decompositions of $\overline{PS}$

The three decompositions of $\overline{PS}$ described above have meaningfully interpretable relationships among one another. These relationships are summarized in Figure 5, adapted from Yates (1981, 1982). It has already been indicated how the relationship between the Sanders and Murphy decompositions arises, through a disaggregation of the Sanders resolution. The relationship between the Murphy and covariance decompositions comes from a similar disaggregation of the reliability-in-the-small. The derivation of this disaggregation is left as an exercise. One of the things Figure 5 makes clear is the nonindependence of resolution and calibration.

4. Properness: A Special Property of the Probability Score

The procedures described above can be put to a number of practical uses. The emphasis in the discussion so far has been upon the forecast user assessing various aspects of forecast quality, perhaps as a guide to the choice of a forecaster whose advice one might follow. Extending the spirit of that perspective, one might also use the procedures introduced to screen and train forecasters and to do research on the forecasting process.

Figure 5.

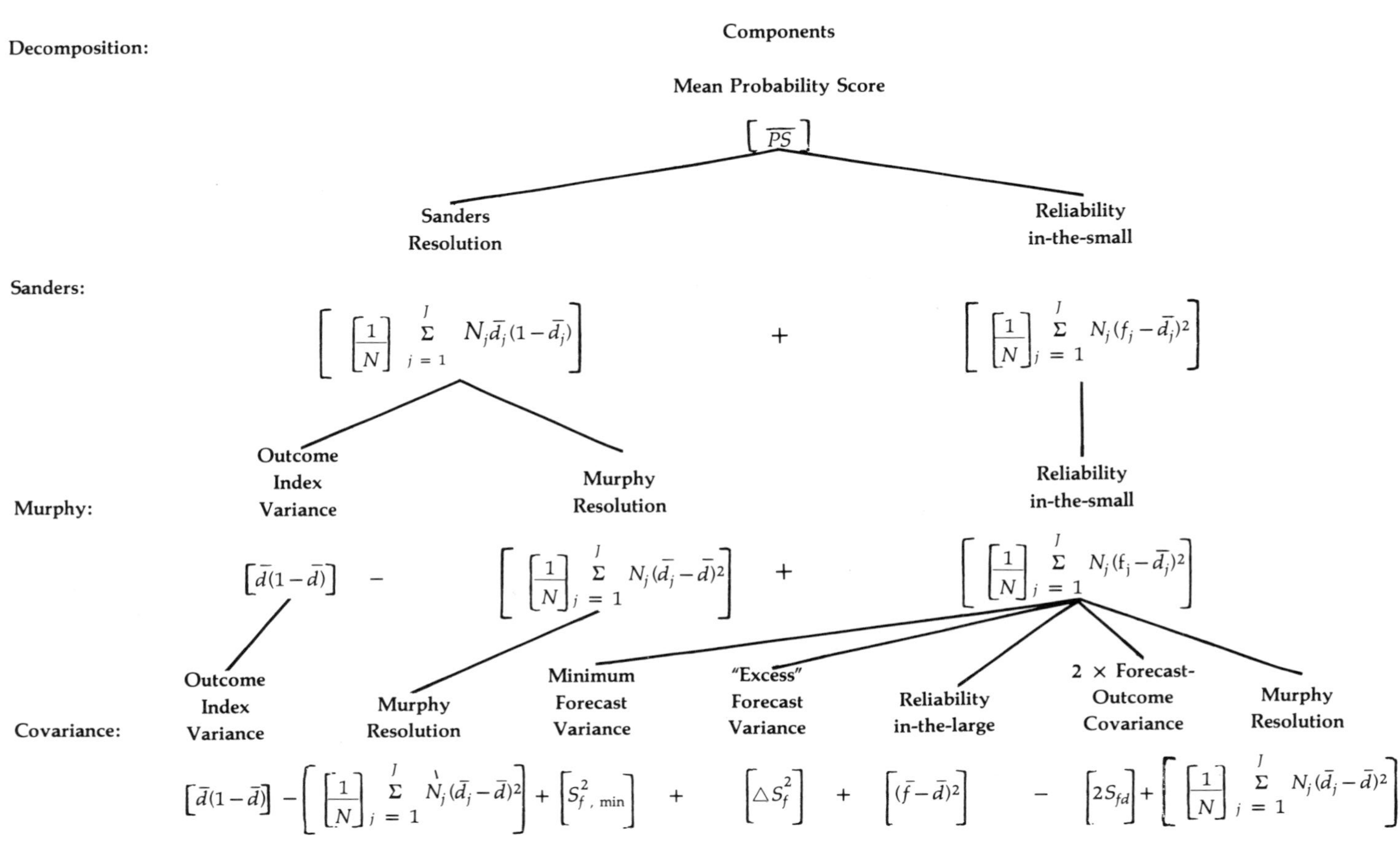
Decomposition:
Components
Mean Probability Score
Sanders Resolution
Reliability in-the-small
Sanders:
Outcome Index Variance
Murphy Resolution
Reliability in-the-small
Murphy:
Outcome Index Variance
Murphy Resolution
Minimum Forecast Variance
"Excess" Forecast Variance
Reliability in-the-large
2 × Forecast-Outcome Covariance
Murphy Resolution
Covariance:

Another potential use of the measures described involves the evaluation of forecasters. The U.S. National Weather Service, for example, grades weather forecasting performance using measures closely related to those presented here. If various scores can be used to *evaluate* good forecasting performance, why not use them to *reward* good performance, too? The probability score *PS* has a property called "properness" (to be described below) that makes it an attractive performance evaluator.

A forecast user naturally wishes to know what a forecaster *really* thinks in a given situation. On the other hand, one can imagine a number of reasons the forecaster might be inclined to "hedge" reported forecasts away from his or her true opinions. For example, the forecaster might believe that the chances of event *A* occurring are extremely good. However, because he or she does not wish to appear foolish in the (perceived) unlikely circumstance that event *A* does not occur, the forecaster offers a moderated, hedged forecast.

Suppose the forecaster is compensated in direct relation to the inverse of *PS*. (Recall that the better external correspondence is, the *lower PS* will be.) As a concrete—even if contrived—example, suppose that graduate admissions counselors' merit pay raises were made in direct relation to the average value of $S(f,d) = 1 - PS(f,d)$, calculated for each admissions forecast f made during the year. Clearly, it is in the counselor's financial interests to minimize $PS(f,d)$.

Let f be the forecaster's "true" probabilistic forecast of event *A*'s occurrence. Suppose $\hat{f} = f + \triangle$ is the forecast the forecaster considers reporting. The report $\hat{f}$ is hedged if $\triangle \neq 0$. Each potential reported forecast can be represented as a gamble $G_{\hat{f}}$, faced by the forecaster. The gamble might be conveniently displayed in tree form:

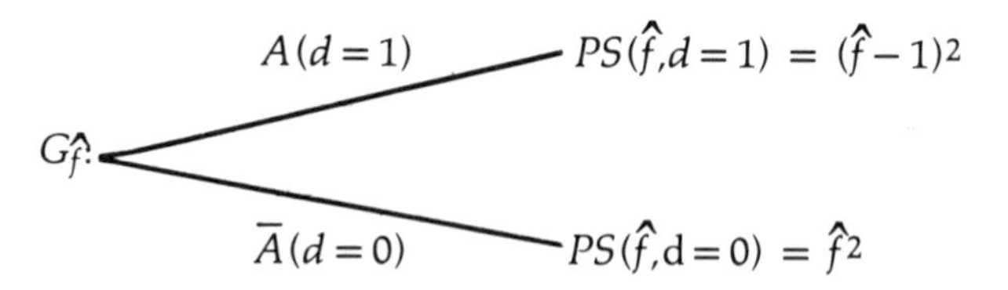

As the tree dramatizes, if the forecaster reports $\hat{f}$ and event *A* occurs, the *PS* value earned is $(\hat{f}-1)^2$; if $\hat{f}$ is reported and event *A* does not occur, the earned *PS* value is $\hat{f}^2$. Now, the expected value of $G_{\hat{f}}$ is given by

$$EV(G_{\hat{f}}) = P(A)(\hat{f}-1)^2 + [1-P(A)]\hat{f}^2. \tag{16}$$

From the forecaster's point of view, $P(A) = f$. Using this fact and $\hat{f} = f + \triangle$, Equation (16) can be reduced to

$$EV(G_{\hat{f}}) = f(1-f) + \triangle^2. \qquad (17)$$

Thus, it is clear that if the forecaster wishes to minimize the expected value of *PS*, he or she should make $\triangle = 0$, i.e., report $\hat{f} = f$, his or her true forecast. Put another way, the forecaster should choose the gamble, i.e., reported forecast $\hat{f}$, that is identical to his or her true forecast f.

Technically and more generally, any scoring rule is said to be *proper* if its expected value is maximized or minimized when the reported forecast is the same as the true forecast. The class of proper scoring rules is not limited to *PS*. However, *PS* is the one most commonly studied and used.

5. Bibliography

Brier, G.W. "Verification of forecasts expressed in terms of probability." *Monthly Weather Review*, 1950, *78*(1), 1-3.

Lichtenstein, S., and Fischhoff, B. "Training for calibration." *Organizational Behavior and Human Performance*, 1980, *26*, 149-171.

Lichtenstein, S., Fischhoff, B., and Phillips, L.D. "Calibration of probabilities: The state of the art to 1980." In D. Kahneman, P. Slovic, and A. Tversky (Eds.), *Judgment Under Uncertainty: Heuristics and Biases*. New York: Cambridge University Press, 1982.

Murphy, A.H. "A new vector partition of the probability score." *Journal of Applied Meteorology*, 1973, *12*, 595-600.

Murphy, A.H., and Winkler, R.L. "Reliability of subjective probability forecasts of precipitation and temperature." *Applied Statistics*, 1977, *26*, 41-47.

Sanders, F. "On subjective probability forecasting." *Journal of Applied Meteorology*, 1963, *2*, 191-201.

Sanders, F. "Skill in forecasting daily temperature and precipitation: Some experimental results." *Bulletin of the American Meteorological Society*, 1973, *54*, 1171-1179.

Shuford, E.H., Jr., Albert, A., and Massengill, H.E. "Admissible probability measurement procedures." *Psychometrika,* 1966, *31,* 125-145.

Staël von Holstein, C.A.S. "Probabilistic forecasting: An experiment related to the stock market." *Organizational Behavior and Human Performance,* 1972, *8,* 139-158.

Winkler, R.L., and Murphy, A.H. "'Good' probability assessors." *Journal of Applied Meteorology,* 1968, *7,* 751-758.

Yates, J.F. "Forecasting Performance: A Covariance Decomposition of the Mean Probability Score." Paper presented at the 22nd Annual Meeting of the Psychonomic Society, Philadelphia, November, 1981.

Yates, J.F. "External correspondence: Decompositions of the mean probability score." *Organizational Behavior and Human Performance,* 1982, *30,* 132-156.

6. Exercises

The following exercises are keyed to pertinent sections of the module, as indicated in parentheses at the end of each exercise statement.

1. Gamble $G1$ offers the player a gain of $9 if a single spot appears on the toss of a fair die, nothing if 2, 3, or 4 spots appear, and a loss of $3 if 5 or 6 spots appear. $G1$ can be represented in "tree" form as follows:

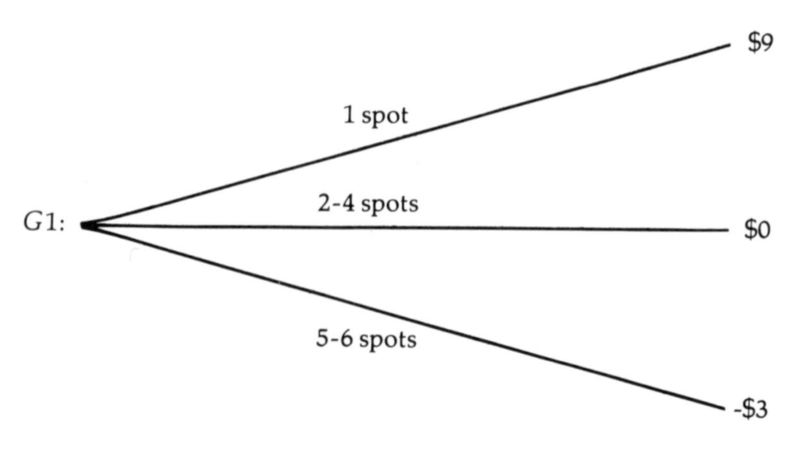

Gamble $G2$ promises a prize of \$8 if heads appear on both of two tosses of a fair coin, a loss of \$1 if a head appears on exactly one of the tosses, and a loss of \$2 if tails appear on both tosses.

a) Compute the expected value (EV) of $G1$.

b) Construct the tree diagram and compute the EV of $G2$.

c) Which gamble would you prefer, $G1$ or $G2$? Which gamble should be chosen according to the expectation maximization principle?

d) Suppose that the coin used in $G2$ is biased, such that $P(H) = .7$ on each toss. What is the "new" EV for $G2$? (1.1)

2. A gamble or prospect is said to be "fair" if its EV is zero.

a) Do you think that an automobile insurance policy is a fair prospect? Defend your position.

b) How about state lottery tickets?

c) What do your reflections about **a**) and **b**) lead you to conclude about the maximization of EV as a theory of how people make decisions about prospects in real life? (1.1)

3. Consider the citrus grower's situation described in Section 1.1. Suppose now that there are additional protection costs beyond the baseline cost C that will be incurred if a freeze actually does occur, e.g., the cost of fueling smudge pots to heat the orchards.

a) Derive a general expression for the cutoff probability for the above conditions.

b) Assuming that $C = -\$1000$ and $L = -\$8000$, as before, and that these additional contingent costs are \$2000, what is the new cutoff probability? (1.1)

4. Show that $B = 0$ in Equation (5). *Hint:* What is

$$\sum_{i=1}^{N_j} (\bar{d}_j - d_{ij})? \text{ (2.4)}$$

5. Show that

$$S^2_{d_{ij}} = (1/N_j) \sum_{i=1}^{N_j} (\bar{d}_j - d_{ij})^2$$

can be expressed as $\bar{d}_j(1-\bar{d}_j)$. (2.4)

6. Derive Equation (6). *Hint:* Define $\triangle_j$ as $\triangle_j = \bar{d}_j - \bar{d}, j = 1 - J$. (2.5)

7. Sports Forecaster 17 was asked to forecast the outcomes of 20 college basketball games. Following are the results of the exercise, where in each ordered pair (f,d), f is Sports Forecaster 17's probabilistic forecast that the home team would win and d is the corresponding outcome index value, 1 when the home team won, 0 when it lost:

(.90,1), (.62,1), (.77,1), (.29,0), (.53,0), (.60,0),
(.50,1), (.42,1), (.80,1), (.44,1), (.56,0), (.50,1),
(.70,1), (.36,1), (.46,0), (.51,1), (.46,1), (.53,1),
(.50,0), (.24,0),

Grade Forecaster B was asked to make probabilistic forecasts that each of 65 students would earn a grade of "B" in a certain course. The following ordered pairs (f,d) are the results, where f is the forecast of a grade of "B," and d is the corresponding outcome index value, 1 when a "B" was earned and 0 when some other grade was earned:

(.00,0), (.70,1), (.60,0), (.10,0), (.40,0), (.00,0),
(.70,0), (.00,0), (.90,1), (.00,0), (.30,0), (.30,1),
(.30,1), (.20,1), (.50,0), (.10,1), (.30,1), (.40,0),
(.50,1), (.20,0), (.30,0), (.20,0), (.20,0), (.10,0),
(.20,1), (.20,1), (.60,1), (.30,1), (.60,0), (.20,1),
(.40,1), (.60,0), (.60,0), (.30,1), (.70,1), (.40,0),
(.70,1), (.10,1), (.50,0), (.00,0), (.70,1), (.40,1),
(.70,0), (.30,0), (.20,0), (.60,1), (.00,0), (.00,0),
(.20,0), (.70,1), (.70,1), (.80,0), (.70,0), (.70,0),
(.40,0), (.80,1), (.40,0), (.30,1), (.40,0), (.80,0),
(.80,1), (.40,1), (.30,1), (.00,0), (.80,0).

Incidentally, the data for Sports Forecaster 17 and Grade Forecaster B are real.

a) For Sports Forecaster 17's data: (**i**) round the forecasts to the nearest tenth; (**ii**) construct the reliability diagram; and (**iii**) calculate $\overline{PS}$ and the components of the Sanders and Murphy decompositions of $\overline{PS}$.

b) As in **a**), using forecast categories of tenths, for Grade Forecaster B's data: (**i**) construct the reliability diagram; and (**ii**) calculate $\overline{PS}$ and the components of the Sanders and Murphy decompositions of $\overline{PS}$.

c) Compare and contrast the forecasting performance of Sports Forecaster 17 and Grade Forecaster B overall and with respect to calibration-in-the-small and resolution. Which would you consider to have better judgment and why? (For purposes of discussion, ignore the smallness of the sample sizes, and thus the likely instability of the statistics.) (2.2 - 2.5)

8. Derive Equation (11). *Hint:* Note that

$f_k - d_k = f_k - \bar{f} + \bar{f} - \bar{d} + \bar{d} - d_k$. (3.3)

9. Derive Equation (13). *Hint:* $r_{fd} = (\bar{f}_1 - \bar{f}_0)\sqrt{N_0}\sqrt{N_1}/S_f N$. (3.3)

10. Show that

$$S^2_{f,\,\min} = (\bar{f}_1 - \bar{f}_0)^2 \bar{d}(1 - \bar{d})$$

is the smallest value of S_f^2 the forecaster can achieve, given particular values of $\bar{f}_1$, $\bar{f}_0$ and $\bar{d}$. (3.3)

11. Derive Equation (15). (3.3)

12. Show that the reliability-in-the-small can be expressed as follows:

$$(1/N) \sum_{j=1}^{J} N_j (f_j - \bar{d}_j)^2 = S_f^2 + (\bar{f} - \bar{d})^2 - 2S_{fd}$$

$$+ (1/N) \sum_{j=1}^{J} N_j (\bar{d}_j - \bar{d})^2.$$

Hint: Observe that $f_j - \bar{d}_j = f_j - \bar{f} + \bar{f} + \bar{d} - \bar{d} - \bar{d}_j$. (3.5)

13. Use the data from Problem 7 for this problem.

a) For Sports Forecaster 17's data: (**i**) construct the covariance graph; and (**ii**) calculate the components of the covariance decomposition of $\overline{PS}$.

b) For Grade Forecaster B's data: (**i**) construct the covariance graph; and (**ii**) calculate the components of the covariance decomposition of $\overline{PS}$.

c) Verify that the component relationships represented by the equation in Problem 12 hold for the data of Sports Forecaster 17 and Grade Forecaster B. (3.2 - 3.5)

14. The *linear scoring rule* is defined as follows:

$$Lr(f,d) = fd + (1-f)(1-d),$$

where f and d are, respectively, the forecast and outcome index for event A. Clearly, $Lr = f$ when event A occurs; $Lr = 1 - f$ when event A does not occur. If Lr is used as a basis for rewarding the forecaster for his or her forecast external correspondence, the forecaster is compensated in direct proportion to the probability assigned to the event that ultimately occurs. On the surface, Lr thus seems to provide a very reasonable means of providing an incentive for accuracy.

a) Show that Lr is not proper.

b) What sort of forecast reporting should the use of Lr as a reward basis encourage?

c) Perform an experiment to test whether the use of PS and Lr *do* in fact lead to different forecasting tendencies. (4)

7. Answers to Exercises

1. a) $EV(G1) = \$.50$
   b) $EV(G2) = \$1.00$
   d) $EV'(G2) = \$3.32$

3. a) $p_c(L - C') = C$, where C' = additional contingent costs.
   b) $p_c = 1/6$

4. Note that

$$\sum_{i=1}^{N_j} (\bar{d}_j - d_{ij}) = \sum_{i=1}^{N_j} \bar{d}_j - \sum_{i=1}^{N_j} d_{ij} = N_j\bar{d}_j - N_j\bar{d}_j = 0.$$

So

$$B = 2(1/N) \sum_{j=1}^{J} (f_j - \bar{d}_j) \sum_{i=1}^{N_j} (\bar{d}_j - d_{ij}) = 0.$$

5. Recall that $d_{ij} = 0$ or 1. Thus $d_{ij}^2 = d_{ij}$. So, after expanding $(\bar{d}_j - d_{ij})^2$, it is clear that

$$S_{d_{ij}}^2 = (1/N_j) \sum_{i=1}^{N_j} \bar{d}_j^2 - 2\bar{d}_j(1/N_j) \sum_{i=1}^{N_j} d_{ij} + (1/N_j) \sum_{i=1}^{N_j} d_{ij}^2$$
$$= \bar{d}_j^2 - 2\bar{d}_j \cdot \bar{d}_j + \bar{d}_j = \bar{d}_j(1 - \bar{d}_j).$$

6. If $\triangle_j = \bar{d}_j - \bar{d}$, then

$$(1/N) \sum_{j=1}^{J} N_j \triangle_j = 0.$$

This is easily seen when it is remembered that

$$\bar{d}_j = (1/N_j) \sum_{i=1}^{N_j} d_{ij} \text{ and}$$

$$\bar{d} = (1/N) \sum_{j=1}^{J} \sum_{i=1}^{N_j} d_{ij}.$$

So, we can write

$$(1/N)\sum_{j=1}^{J} N_j\bar{d}_j(1-\bar{d}_j) = (1/N)\sum_{j=1}^{J} N_j[(\triangle_j+\bar{d}) - (\triangle_j+\bar{d})^2]$$

$$= (1/N)\sum_{j=1}^{J} N_j\triangle_j + (1/N)\sum_{j=1}^{J} N_j\bar{d}$$

$$-(1/N)\sum_{j=1}^{J} N_j(\triangle_j+\bar{d})^2$$

$$= 0 + \bar{d} - (1/N)\sum_{j=1}^{J} N_j(\triangle_j^2 + 2\triangle_j\bar{d} + \bar{d}^2)$$

$$= \bar{d}(1-\bar{d}) - (1/N)\sum_{j=1}^{J} N_j(\bar{d}_j-\bar{d})^2,$$

after terms are collected and $\triangle_j$ is replaced by $\bar{d}_j - \bar{d}$.

7. **a)** **(iii)** $\overline{PS} = .2135$
$SR = .1271$
$RIS = .0864$
$S_d^2 = .2275$
$MR = .1004$

b) **(ii)** $\overline{PS} = .2640$
$SR = .1988$
$RIS = .0652$
$S_d^2 = .2452$
$MR = .0464$

8. $f_k - d_k = (f_k-\bar{f}) + (\bar{f}-\bar{d}) + (\bar{d}-d_k)$ implies that

$$(1/N)\sum_{k=1}^{N} (f_k-d_k)^2$$

can be expanded as follows:

$$(1/N)\sum_{k=1}^{N} (f_k-d_k)^2 = A + B + C + D + E + F, \text{ where}$$

$$A = (1/N)\sum_{k=1}^{N} (f_k-\bar{f})^2 = S_f^2,$$

$$B = (1/N)\sum_{k=1}^{N} 2(f_k-\bar{f})(\bar{f}-\bar{d}) = 2(\bar{f}-\bar{d})(1/N)\sum_{k=1}^{N} (f_k-\bar{f}) = 0,$$

$$C = (1/N) \sum_{k=1}^{N} 2(f_k - \bar{f})(\bar{d} - d_k) = -2(1/N) \sum_{k=1}^{N} (f_k - \bar{f})(d_k - \bar{d}) = -2S_{fd},$$

$$D = (1/N) \sum_{k=1}^{N} (\bar{f} - \bar{d})^2 = (\bar{f} - \bar{d})^2,$$

$$E = (1/N) \sum_{k=1}^{N} 2(\bar{f} - \bar{d})(\bar{d} - d_k) = 2(\bar{f} - \bar{d})(1/N) \sum_{k=1}^{N} (\bar{d} - d_k) = 0, \text{ and}$$

$$F = (1/N) \sum_{k=1}^{N} (\bar{d} - d_k)^2 = S_d^2.$$

9. Since d is a dichotomous variable, r_{fd} is a point-biserial correlation coefficient and can be written as $r_{fd} = (\bar{f}_1 - \bar{f}_0)\sqrt{N_0}\sqrt{N_1}/S_f N$. Also, since N_1 is the total number of times that d assumes the value 1, $\bar{d} = N_1/N$. Similarly, it can be seen that $(1-\bar{d}) = N_0/N$. So, $S_d^2 = \bar{d}(1-\bar{d}) = N_0 N_1/N^2$. Making the implied substitutions, we find that $S_{fd} = r_{fd} S_f S_d = (\bar{f}_1 - \bar{f}_0)\bar{d}(1-\bar{d})$.

10. Let f_{1m}, $m = 1 - N_1$, represent the values of f on the N_1 occasions when $d = 1$, and let f_{0n}, $n = 1 - N_0$, represent the values of f on the remaining N_0 occasions when $d = 0$. Then the variance of f can be written as

$$S_f^2 = (1/N) \sum_{k=1}^{N} (f_k - \bar{f})^2$$

$$= (1/N) \sum_{m=1}^{N_1} [(f_{m1} - \bar{f}_1) + (\bar{f}_1 - \bar{f})]^2$$

$$+ (1/N) \sum_{n=1}^{N_0} [(f_{n0} - \bar{f}_0) + (\bar{f}_0 - \bar{f})]^2.$$

Now

$$(1/N) \sum_{m=1}^{N_1} [(f_{m1}-\bar{f}_1) + (\bar{f}_1-\bar{f})]^2$$

$$= (1/N) \sum_{m=1}^{N_1} (f_{m1}-\bar{f}_1)^2 + 2(\bar{f}_1-\bar{f})(1/N) \sum_{m=1}^{N_1} (f_{m1}-\bar{f}_1) + (1/N) \sum_{m=1}^{N_1} (\bar{f}_1-\bar{f})^2$$

$$= (1/N) \sum_{m=1}^{N_1} (f_{m1}-\bar{f}_1)^2 + (N_1/N)(\bar{f}_1-\bar{f})^2,$$

since $\sum_{m=1}^{N_1} (f_{m1}-\bar{f}_1) = 0.$

Similarly, it can be shown that

$$(1/N) \sum_{n=1}^{N_0} [(f_{n0}-\bar{f}_0) + (\bar{f}_0-\bar{f})]^2 = (1/N) \sum_{n=1}^{N_0} (f_{n0}-\bar{f}_0)^2 + (N_0/N)(\bar{f}_0-\bar{f})^2.$$

Recognizing that

$\bar{f} = [N_1 \bar{f}_1 + N_0 \bar{f}_0]/N$, $\bar{d} = N_1/N$, and $(1-\bar{d}) = N_0/N$, we see that

$$\textbf{(A)} \quad S_f^2 = (1/N)\left[\sum_{m=1}^{N_1} (f_{m1}-\bar{f}_1)^2 + \sum_{n=1}^{N_0} (f_{n0}-\bar{f}_0)^2\right] + (\bar{f}_1-\bar{f}_0)^2\bar{d}(1-\bar{d}).$$

Since $(f_{m1}-\bar{f}_1)^2$, $m = 1 - N_1$, and $(f_{n0}-\bar{f}_0)^2$, $n = 1 - N_0$, must be non-negative, the minimum value of S_f^2 is

$$S_{f,\,\min}^2 = (\bar{f}_1-\bar{f}_0)^2\bar{d}(1-\bar{d}).$$

11. From Equation (A) in the above solution to Exercise 10,

$$S_f^2 - S_{f,\,\min}^2 = (1/N)\left[\sum_{m=1}^{N_1} (f_{m1}-\bar{f}_1)^2 + \sum_{n=1}^{N_0} (f_{n0}-\bar{f}_0)^2\right].$$

The result follows from the definitions of $S_{f_0}^2$ and $S_{f_1}^2$.

12. Note that $f_j - \bar{d}_j$ can be expressed as

$$f_j - \bar{d}_j = (f_j-\bar{f}) + (\bar{f}-\bar{d}) + (\bar{d}-\bar{d}_j).$$

We can then write

$$(1/N) \sum_{j=1}^{J} N_j(f_j-\bar{d}_j)^2 = A + B + C + D + E + F, \text{ where}$$

$$A = (1/N) \sum_{j=1}^{J} N_j(f_j-\bar{f})^2 = S_f^2,$$

$$B = (1/N) \sum_{j=1}^{J} 2N_j(f_j-\bar{f})(\bar{f}-\bar{d}) = 2(\bar{f}-\bar{d})(1/N) \sum_{j=1}^{J} N_j(f_j-\bar{f}) = 0,$$

$$C = (1/N) \sum_{j=1}^{J} 2N_j(f_j-\bar{f})(\bar{d}-\bar{d}_j) = -2(1/N) \sum_{j=1}^{J} N_j(f_j-\bar{f})(\bar{d}_j-\bar{d}) = -2S_{fd},$$

$$D = (1/N) \sum_{j=1}^{J} N_j(\bar{f}-\bar{d})^2 = (\bar{f}-\bar{d})^2,$$

$$E = (1/N) \sum_{j=1}^{J} 2N_j(\bar{f}-\bar{d})(\bar{d}-\bar{d}_j) = 2(\bar{f}-\bar{d})(1/N) \sum_{j=1}^{J} N_j(\bar{d}-\bar{d}_j) = 0, \text{ and}$$

$$F = (1/N) \sum_{j=1}^{J} N_j(\bar{d}-\bar{d}_j)^2.$$

13. a) (ii) $\overline{PS} = .2104$
Bias $= -.116$
Slope $= .123$
$\triangle S_f^2 = .0222$

b) (ii) $\overline{PS} = .2640$
Bias $= -.0345$
Slope $= .100$
$\triangle S_f^2 = .0640$

14. a) Let f be the forecaster's "true" forecast and let $\hat{f} = f + \triangle$ be a forecast he or she considers reporting. The expected value of reporting $\hat{f}$ can be written as

(B) $$EV(G_{\hat{f}}) = f \cdot \hat{f} + (1-f)(1-\hat{f}) = 2f^2 - 2f + 1 + (2f-1)\triangle.$$

Clearly, if $(2f-1) > 0$, $EV(G_{\hat{f}})$ is not maximized when $\triangle = 0$, i.e., when the forecaster reports his or her true forecast. So, the linear scoring rule is not proper.

b) Equation (**B**) above implies that, if the forecaster thinks event A is more likely than not to occur, then he or she should report $\hat{f} = 1.0$. If event A is thought to be less likely than not, then the forecaster should report $\hat{f} = 0.0$.

UMAP

Modules in Undergraduate Mathematics and its Applications

Module 634

Probability in a Contested Election

Dennis C. Gilliland

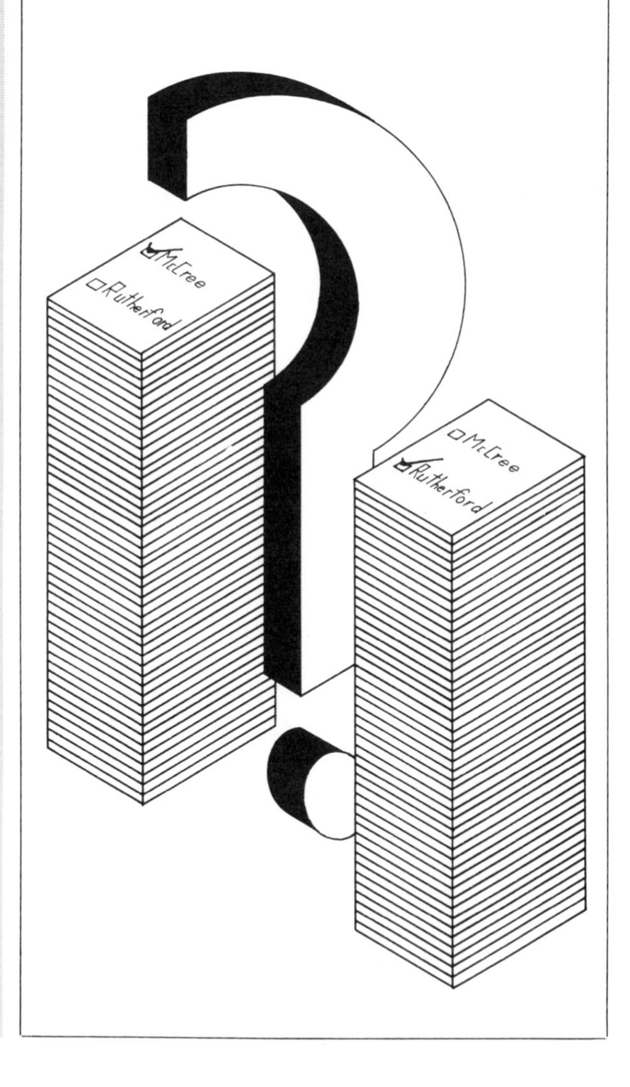

Published in cooperation with the Society for Industrial and Applied Mathematics, the Mathematical Association of America, the National Council of Teachers of Mathematics, the American Mathematical Association of Two-Year Colleges, and The Institute of Management Sciences.

INTERMODULAR DESCRIPTION SHEET: UMAP Unit 634

TITLE: PROBABILITY IN A CONTESTED ELECTION

AUTHOR: Dennis C. Gilliland
Department of Statistics and Probability
Michigan State University
East Lansing, Michigan 48824

MATH FIELD: Probability and statistics

APPLICATIONS FIELD: Political science, law

ABSTRACT: Elementary probability models can sometimes be used to analyze the results of an election where irregularities have occurred. In this module we give an illustrative example by considering the 1975 mayoral election in the city of Flint, Michigan. True vote is estimated with an estimator which is used in randomized response models. Students learn an application of a simple discrete probability model, and reinforce knowledge concerning mean, standard deviation, the binomial distribution, independence, and convolution.

PREREQUISITES: Knowledge of discrete probability, including mean, standard deviation and the binomial. Method of moments, convolutions, and the normal distribution.

The UMAP Journal, Vol. V, No. 1, 1984

Probability in a Contested Election

Dennis C. Gilliland
Department of Statistics and Probability
Michigan State University
East Lansing, Michigan 48824

Table of Contents

1. INTRODUCTION 1
2. FACTS CONCERNING THE FLINT ELECTION 1
3. A MODEL TO ESTIMATE THE TRUE VOTE 3
4. CRITIQUE AND COMMENTS 5
5. RESOLUTION OF THE PRESENT CASE 6
6. EXERCISES 7
7. REFERENCES 8
8. SOLUTIONS TO EXERCISES 9

Modules and Monographs in Undergraduate Mathematics and its Applications Project (UMAP)

The goal of UMAP was to develop, through a community of users and developers, a system of instructional modules in undergraduate mathematics and its applications that may be used to supplement existing courses and from which complete courses may eventually be built.

The Project was guided by a National Advisory Board of mathematicians, scientists, and educators. UMAP was funded by a grant from the National Science Foundation to the Consortium for Mathematics and Its Applications, Inc. (COMAP), a nonprofit corporation engaged in research and development in mathematics education.

UMAP Advisory Board

The Project would like to thank James Inglis of Bell Laboratories and Gary Simon of New York University for their reviews, and all others who assisted in the production of this unit.

1. Introduction

The result of a close election may be challenged on the basis of irregularities and the outcome may ultimately be decided by the courts. A given court may uphold the result, reverse the result, invalidate the entire election, or offer any of a number of remedies. In general, an election will not be overturned on the basis of a mere mathematical *possibility* that the results would be reversed in the absence of irregularities. Being reluctant to disenfranchise the valid electors unjustly, the courts have sometimes required that a challenger establish a *probability* that the result would have been reversed in the absence of irregularities.

In this unit we discuss the contested mayoral election held in the city of Flint, Michigan on November 4, 1975 and a model used to quantify the probability of reversal.

2. Facts Concerning the Flint Election

The city of Flint, Michigan held an election on November 4, 1975 for the office of Mayor, wherein a mayor was to be elected for the first time under a new city charter. The vote for mayor after the official canvass showed James Rutherford the winner over Floyd McCree with a margin of 206 votes (Table 1).

Table 1

Recorded Vote, 1975

Precinct	Candidate McCree	Rutherford	Totals
51	202	253	455
52	174	117	291
Totals (51 and 52)	376	370	746
Other	20,099	20,311	40,410
Totals	20,475	20,681	41,156

The breakdown for Precincts 51 and 52 is given because these vote totals were disputed due to a mix-up in the voting devices in these precincts. The vote totals from the other 143 precincts and absentee voters were not disputed.

Voting was done in Flint with punch cards. Here each voter at a precinct received a punch card which he or she took to a voting booth. In the booth was a voting device that worked as follows. The punch card was positioned to be punched by placing it in a slot. A booklet, which was fixed to the voting device, listed for each office the names of the candidates, with a hole beside each name. The individual voted for the candidate of choice by punching through the hole with the stylus that was provided. This removed a square (chad) from the punch card. All persons voting at a given precinct deposited their punch cards in the same receptacle as they were leaving the room in which the voting devices were located. The votes were subsequently counted by computer.

Precinct 51 in Flint had five voting devices and Precinct 52 had four voting devices. Rutherford was to be listed first among the two candidates for mayor in each of the booklets placed in the voting devices of Precinct 51. McCree was to be listed first in each of the booklets placed in the voting devices of Precinct 52.

By mistake, the election officials placed one booklet in Precinct 51 that should have been in Precinct 52, and one booklet in Precinct 52 that should have been in Precinct 51. The result was that each vote cast in the voting booth with the wrong booklet in either precinct was reversed, that is, recorded for the other candidate. The commingling of punch cards from all voting devices within a precinct made it impossible to distinguish which or how many votes were reversed within Precincts 51 and 52. The error was discovered after the polls had closed and the devices disassembled so that not even the locations within the polling places of the voting booths with reversed booklets were known.

A total of 746 votes were cast in Precincts 51 and 52 of which 376 were recorded for McCree. Because of the mix-up in booklets, it is possible that the total vote cast for McCree in these precincts was at least the 479 necessary to overcome the undisputed lead of 212 that Rutherford had across the other precincts (see Exercise 1). On this basis, Floyd McCree challenged in the courts the decision by the Board of Canvassers to declare James Rutherford the winner.

In the next section we present a model that was presented in court and used to assess the likelihood that, in fact, the number of votes cast for McCree in Precincts 51 and 52 was sufficient for him to have been the winner.

3. A Model to Estimate the True Vote

To estimate the true vote for a precinct that has a reversed booklet, we use the following model. Let k denote the number of voting booths in the precinct, of which exactly one has a reversed ballot booklet, and let $p = 1/k$. Suppose that θ votes are cast for McCree and $n - \theta$ votes are cast for Rutherford in the precinct. The number m recorded for McCree is given by

$$m = x + y, \tag{3.1}$$

where x is the number of votes cast for McCree on the $k - 1$ proper booklets and y is the number of votes cast for Rutherford but mistakenly recorded for McCree on the reversed booklet.

Under uniform random selection of voting booths by voters, one would expect $k - 1$ out of every k votes cast for McCree to be recorded for McCree and 1 out of every k votes cast for Rutherford to be recorded for McCree. Thus, the expected value of x is

$$E(x) = \theta(1-p),$$

the expected value of y is

$$E(y) = (n-\theta)p,$$

and the expected value of m is

$$E(m) = \theta(1-p) + (n-\theta)p. \tag{3.2}$$

Using the method of moments to derive an estimator of θ, we equate m and the right-hand side of (3.2). Provided $p \neq 1/2$, that is, $k > 2$, the resulting equation can be solved for θ, to yield the unbiased estimator

$$\hat{\theta} = \frac{m - np}{1 - 2p} \cdot \tag{3.3}$$

The estimator $\hat{\theta}$ is a function of the random variable m, which is the sum of the random variables x and y. The probability distribution of $\hat{\theta}$ is important in assessing the accuracy of $\hat{\theta}$. It can be determined under a variety of assumptions on the way voters select voting booths. We consider one such assumption.

ASSUMPTION: Within a disputed precinct, voters independently select the voting booths in which they cast their ballots. For each voter, the probability is $p \neq 1/2$ that the voting booth with the reversed ballot booklet is selected.

Under this assumption, the number of votes cast for McCree and recorded for McCree, x, can be thought of as the number of heads in θ independent tosses of a coin with probability $1 - p$ of a head on a single toss. Its probability distribution is binomial with parameters θ and $1 - p$, denoted by $x \simeq B(\theta, 1-p)$. Similarly, $y \simeq B(n-\theta, p)$ independently of x, and the probability distribution of $m = x + y$ is the convolution of these two binomial distributions, i.e.,

$$m \simeq B(\theta, 1-p) * B(n-\theta, p), \qquad \theta \in T, \tag{3.4}$$

where $T = \{0, 1, \ldots, n\}$. The mean of a binomial distribution is the number of trials times the probability of success on a single trial, so (3.2) holds. Using the formula for the variance of a binomial, we find m has variance

$$\sigma^2(m) = \theta(1-p)p + (n-\theta)p(1-p) = np(1-p). \tag{3.5}$$

Thus, under our assumption, $\hat{\theta}$ of (3.3) is an unbiased estimator for θ with variance

$$\sigma^2(\hat{\theta}) = \frac{np(1-p)}{(1-2p)^2}. \tag{3.6}$$

The estimator $\hat{\theta}$ has the nice property that its variance does not depend on the parameter θ.

Table 2 gives the results of applying (3.3) and the data of Table 1 to estimate the true vote for McCree in Precinct 51 ($p = 1/5$) and Precinct 52 ($p = 1/4$). The estimates (in Table 2) for Rutherford are determined by $n - \hat{\theta}$ in each precinct. Using properties of expectation and standard deviation, $E(n-\hat{\theta}) = n - \theta$ and $\sigma(n-\hat{\theta}) = \sigma(\hat{\theta})$. Standard deviations are indicated in parentheses beside each estimate in Table 2.

The standard deviation found in the bottom row comes from the formula for the standard deviation of the sum of two independent random variables, namely,

$$\sigma(\hat{\theta}_1 + \hat{\theta}_2) = (\sigma^2(\hat{\theta}_1) + \sigma^2(\hat{\theta}_2))^{1/2}$$
$$= (14.2^2 + 14.7^2)^{1/2}$$
$$= 20.5.$$

Table 2

Estimates of True Vote, 1975

Precinct	Candidate McCree	Candidate Rutherford	Totals n
51	185 (14.2)	270 (14.2)	455
52	202.5 (14.7)	88.5 (14.7)	291
Totals	387.5 (20.5)	358.5 (20.5)	746

Of course, here we are correctly taking x and y for Precinct 51 to be independent of x and y for Precinct 52, from which the separate estimators $\hat{\theta}_1$ for Precinct 51 and $\hat{\theta}_2$ for Precinct 52 are independent.

The estimated total for McCree, 387.5, and its standard deviation 20.5 indicate that, in all probability, McCree did not have a total $\theta_1 + \theta_2 \geq 479$ sufficient to overcome the undisputed margin of 212 votes Rutherford held outside these precincts. Since $\hat{\theta}_1 + \hat{\theta}_2$ is approximately normally distributed and 479 is 4.5 standard deviations from 387.5, the evidence is overwhelming in favor of this conclusion.

4. Critique and Comments

Our assumption may be attacked as unrealistic because it says that voters select voting booths independently. During a busy period on election day, the voting booth selected by a voter does influence the choice of voting booth by other voters. In the present case, the voter turnout was considered relatively light so that this is not a serious criticism of the model in its intended application.

The estimates derived from the model are very sensitive to the specification p. In the present case, there was no testimony accepted in court concerning the physical locations within each polling place of the voting devices with the reversed booklets. The specification $p = 1/k$ is a neutral or symmetric position relative to the evidence. A mixture model approach that regards p as following a distribution of possible values has certain merits but is beyond the scope of this discussion.

5. Resolution of the Present Case

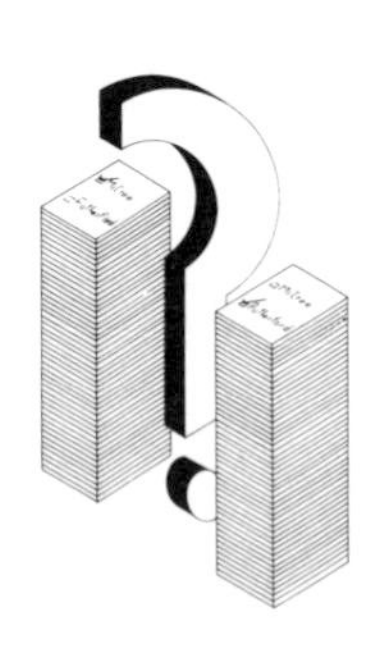

The Board of Canvassers declared Rutherford the winner of the November 4, 1975 election on the basis of the 206-vote margin reported in the bottom line of Table 1. The ensuing court case was heard in Circuit Court of the County of Genesee on February 20, 1976. Plaintiffs presented models with certain assumptions to show that McCree was the actual winner of the November election with a likelihood as great as 1 in 5. Plaintiffs sought relief in the form of a new citywide election. Defendants presented a probability analysis similar to that discussed in this article in an attempt to establish that Rutherford was the actual winner in all probability. They argued that if relief should be given, it should be in the form of a special "mini-election" restricted to the voters who voted in Precincts 51 and 52 in November.

On March 4, the Court ruled in favor of the special mini-election and in the Order indicated that "those voters are instructed to cast their ballot in the same way" (as cast in November). The letter to individual voters phrased it, "You are instructed to recast your ballot at this rerun election so that we may ascertain how you voted on November 4, 1975."

We hoped that the mini-election of April 13, 1976 would reveal the true vote of November 4, 1975, so that the Table 2 estimates could be compared to the actual votes. Many factors caused this hope to be unrealized. To begin with, 763 voters voted in person at Precincts 51 and 52 in November, of whom only 746 cast votes in the mayoral election. All 763 were by necessity eligible to vote in the rerun. Not all people eligible to vote in the rerun bothered to or were around to participate in the April election, as Table 3 shows.

Table 3

Results of April 13, 1976 Mini-Election

Precinct	Candidate McCree	Rutherford	Totals
51	122	255	377
52	217	35	252
Absentee Voters	10	22	32
Totals	349	312	661

The results of the voting by absentee voters were not kept separate for the two precincts. In addition to the 661 votes, seven ballots were cast with no vote shown for either candiate.

The polarization evident in Table 3 may be explained by behind-the-scenes campaigning in the two precincts, which was not explicitly prohibited by the Court Order. On the basis of demographic characteristics and voting patterns, and ignoring the recorded vote for Precincts 51 and 52 in November 4, 1975 (Table 1), it was fairly clear before the rerun mini-election that Precinct 51 favored Rutherford and Precinct 52 favored McCree. The shift to the favorites within the two precincts in excess of what was suggested by the model and Table 2 may be attributed to behind-the-scenes campaigning before the mini-election. Of course, the discrepancy between Tables 2 and 3 can also be explained by the conjecture that the probability p of using the booth with the reversed booklet was greater than 1/5 in Precinct 51 and greater than 1/4 in Precinct 52; i.e., that the booths with the reversed booklets were in positions to attract a larger proportion of the voters than assumed in the model.

In any event, the net effect of the rerun mini-election was to establish Rutherford "firmly" in office with a citywide margin of 175 votes in a total of over 41,000 votes cast.

6. Exercises

1. Verify that McCree ties or beats Rutherford in the November 4, 1975 election if and only if $\theta_1 + \theta_2 \geq 479$.

2. Under the *Assumption*, $\hat{\theta}_1 + \hat{\theta}_2$ is approximately normally distributed with mean $\theta_1 + \theta_2$ and standard deviation $(\sigma^2(\hat{\theta}_1) + \sigma^2(\hat{\theta}_2))^{1/2}$.

   a) Use this information, (3.6) and the specification $p_1 = 0.20$, $p_2 = 0.25$ to construct an approximate 95% confidence interval estimate for $\theta_1 + \theta_2$.

   b) Repeat (a) but with the specification $p_1 = 0.10$ and $p_2 = 0.35$ to demonstrate the sensitivity of the confidence interval estimate to the specification of p_1 and p_2.

3. Show that the estimator (3.3) can take values outside the set T of possible values of θ and even negative values. How can $\hat{\theta}$ be improved?

4. Suppose that the irregularity in the Flint election were not the one indicated, but as follows. Suppose that $N = 400$ illegal votes were included in the total of 41,156 votes with candidates R and M having respective totals 20,681 and 20,475. Suppose there is no evidence about how the illegal votes were cast.

   a) If X denotes the number of illegal votes cast for R, show that the election is tied or McCree is the rightful winner if and only if $X \geq 303$.

   b) In the model which regards X as the number of votes for R in a random sample of $N = 400$ taken from the 41,156 votes, what is the distribution of X? What is the mean of X? What is the standard deviation of X?

7. References

"Developments in the Law-Elections," (1975), *Harvard Law Review 88*, 1111-1339. (Contains a comprehensive review of election law, including postelection remedies.)

Downs, T., Gilliland, D.C. and Katz, L. (1978), "Probability in a Contested Election," *American Statistician, 32*, 122-125.

Finkelstein, M.O. and Robbins, H.E. (1973), "Mathematical Probability in Election Challenges," *Columbia Law Review, 73*, 241-248.

Gilliland, D.C. and Meier, P. (1985), "On the Probability of Reversal in Contested Elections," *Statistics and the Law*, DeGroot, Fienberg and Kadane, Editors. Wiley, New York.

Ward, III., J.E. (1981), "The Probability of Election Reversal," *Mathematics Magazine, 5*, 256-259.

8. Solutions to Exercises

1. Rutherford has an undisputed margin of 212 votes from precincts outside Precincts 51 and 52. Therefore, the vote total $\theta_1 + \theta_2$ for McCree in Precincts 51 and 52 must be at least equal to $(n_1 - \theta_1) + (n_2 - \theta_2) + 212$ in order for McCree to tie or beat Rutherford across all precincts. Here $n_1 = 455$ and $n_2 = 291$ and $\theta_1 + \theta_2 \geq (455 - \theta_1) + (291 - \theta_2) + 212$ if and only if $\theta_1 + \theta_2 \geq 479$.

2. a)
$$\hat{\theta}_1 = [202 - 455(.2)]/[1 - 2(.2)] = 185,$$
$$\sigma^2(\hat{\theta}_1) = 202.22$$
and
$$\hat{\theta}_2 = [174 - 291(.25)]/[1 - 2(.25)] = 202.5,$$
$$\sigma^2(\hat{\theta}_1) = 218.25$$
so that
$$\hat{\theta}_1 + \hat{\theta}_2 = 387.5$$
and
$$\sigma(\hat{\theta}) = 20.5$$
and an approximate 95% *CI* for $\theta_1 + \theta_2$ is
$$387.5 \pm 2\sigma(\hat{\theta}) = 387.5 \pm 41.0.$$

b)
$$\hat{\theta}_1 = 195.6, \ \sigma^2(\hat{\theta}_1) = 63.98;$$
$$\hat{\theta}_2 = 240.5, \ \sigma^2(\hat{\theta}_2) = 735.58$$
so that an approximate 95% *CI* for $\theta_1 + \theta_2$ is 436.1 ± 56.6.

3. $T = \{0, 1, \ldots, n\}$ and $\hat{\theta}$ is not integer valued as a function of the observed total m for McCree. If $m < np$, then $\hat{\theta} < 0$. Replacing $\hat{\theta}$ by $\hat{\theta}*$ defined to be a T-valued minimizer of the distance from $\hat{\theta}$ to T can only reduce the error of estimation. The maximum likelihood estimator of θ has not been determined.

4. **a)** $20{,}475 - (400 - X) \geq 20{,}681 - X$ if and only if $X \geq 303$.

b) X has a hypergeometric distribution with mean $400(20{,}681/41{,}156) = 201.00$ and standard deviation 9.95. X is approximately binomially distributed with $N = 400$ trials, $p = 20{,}681/41{,}156 = 0.5025$; and X is approximately normally distributed as well. See Finkelstein and Robbins (1973) for a detailed discussion of the model which quantifies the probability of reversal by this approach when the irregularity is as indicated. Ward (1981) gives additional examples. Gilliland and Meier (1985) critique the approach.

UMAP

Modules in Undergraduate Mathematics and its Applications

Module 641

Cassette Tapes: Predicting Recording Time

Arnold J. Insel

Published in cooperation with the Society for Industrial and Applied Mathematics, the Mathematical Association of America, the National Council of Teachers of Mathematics, the American Mathematical Association of Two-Year Colleges, and The Institute of Management Sciences.

COMAP

INTERMODULAR DESCRIPTION SHEET:	UMAP Unit 641
TITLE:	CASSETTE TAPES: PREDICTING RECORDING TIME
AUTHOR:	Arnold J. Insel Department of Mathematics Illinois State University Normal, Illinois 61761
MATH FIELD:	Intermediate algebra
APPLICATION FIELD:	Practical experience
TARGET AUDIENCE:	Students enrolled in intermediate algebra course.
ABSTRACT:	The techniques of elementary algebra and geometry are applied to solve the problem of relating the recording time of a cassette tape to the numerical reading of the mechanical counter of a cassette tape deck.
PREREQUISITES:	Familiarity with systems of linear equations in two unknowns and the quadratic formula.

The UMAP Journal, Vol. V, No. 2, 1984

Cassette Tapes: Predicting Recording Time

Arnold J. Insel
Department of Mathematics
Illinois State University
Normal, Illinois 61761

Table of Contents

1. INTRODUCTION 1
2. MATHEMATICAL SOLUTION 2
   2.1 Mechanical Relationship Between the Counter and Tape 2
   2.2 Solution to the Problem 3
3. PRACTICAL COMPUTATIONS 7
   3.1 Predicting Recording Time 7
   3.2 The Inverse Relation 9
4. ANSWERS TO EXERCISES 12

Modules and Monographs in Undergraduate Mathematics and its Applications Project (UMAP)

The goal of UMAP was to develop, through a community of users and developers, a system of instructional modules in undergraduate mathematics and its applications that may be used to supplement existing courses and from which complete courses may eventually be built.

The Project was guided by a National Advisory Board of mathematicians, scientists, and educators. UMAP was funded by a grant from the National Science Foundation to the Consortium for Mathematics and its Applications, Inc. (COMAP), a nonprofit corporation engaged in research and development in mathematics education.

The Project would like to thank Kerry Bailey of Pikes Peak Community College, Harold Baker of Litchfield High School, Patrick Boyle of Santa Rosa Junior College, and Joel Greenstein of New York City Technical College for their reviews, and all others who assisted in the production of this unit.

1. Introduction

A stereo cassette tape deck is a useful component for a hi-fidelity sound system. It records a stereo output from an FM radio or a record onto a magnetic tape sealed into a plastic case called a cassette. Cassettes are rated according to the total number of minutes they can record on both sides. For example, a 90-minute cassette will record for 45 minutes on each of two sides.

If you have used a portion of one side of a cassette to record, it is useful to know how much time remains on that side for additional recording. To aid in this, the tape deck is provided with a tape counter and a reset button that resets the counter to zero. As the tape in the cassette moves forward, the counter measures the amount of tape used. Unfortunately there is no simple relationship between the amount of tape used and the counter reading. The following table gives actual recording times and counter values as obtained on the author's cassette tape deck.

Table 1

time in minutes	counter reading
1	30
2	60
3	88
4	115
5	141
10	262
15	369
20	466
25	556
30	640

Notice that the counter reads 30 at the end of one minute but 262, rather than 300, at the end of ten minutes.

The purpose of this module is to apply elementary algebra and geometry to discover the relationship between counter readings and recording times.

2. Mathematical Solution

2.1 Mechanical Relationship Between Counter and Tape Motion

Figure 1 shows how the tape in a cassette moves during recording or playback. The tape unwinds from the left spool, travels down and across the bottom of the cassette, passes the erasing, recording, and playback heads of the tape deck, and then goes up and winds around the right spool. The tape moves past the heads at the constant rate of one and three quarters inches per second or one hundred five inches per minute.

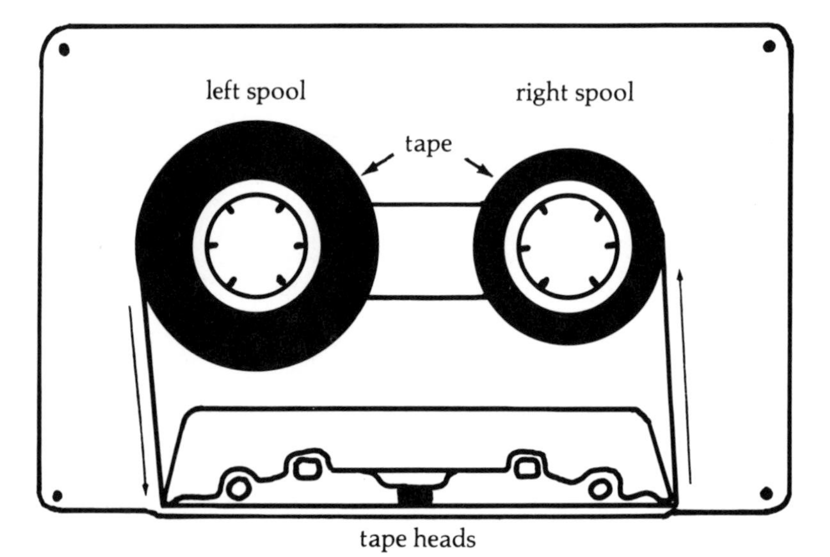

Figure 1. Inside a tape cassette

If we compare the motion within the cassette with the motion of the counter, we discover that the counter measures the turning of the right spool rather than the motion of the tape itself. Suppose we reset the counter to zero and allow the cassette to turn forward. For k equal to the number of revolutions of the right spool and n equal to the counter reading, the following table of values relating k and n was obtained by experimenting with the author's tape deck.

Table 2

k = number of revolutions of right spool	n = counter reading
4	3
8	6
12	9
16	12
20	15
24	18

As n doubles, so does k. More precisely, the ratio k/n is a constant independent of the values of k and n. When two quantities are related so that their ratio is constant, we say that one of the quantities is *directly proportional* to the other one. If c is the value of the constant ratio k/n we have

$$k = cn. \tag{1}$$

In this example $c = 4/3$.

There is another relationship that is useful. Let T be the number of minutes that the tape has been turning forward and let L be the length in inches of the tape wound around the right spool. Since the tape moves forward at the rate of 105 inches per minute,

$$L = 105T$$

$$\text{or } T = L/105. \tag{2}$$

In the next section we shall discover a mathematical relationship between T and n.

2.2 Solution to the Problem

Recall from Section 2.1 that

k is the number of revolutions of the right spool.
n is the counter reading.
L is the length of tape, in inches, wound around the right spool.
T is elapsed time, in minutes.

These variables are related by the equations

$$k = cn \tag{1}$$

$$\text{and } T = L/105. \tag{2}$$

Equations (1) and (2) establish relationships between k and n and between T and L. If an equation relating L and k can be found, it can be used to derive a formula relating T and n.

Since the spool is circular, we can use the familiar geometrical fact that a circle of radius r has circumference $2\pi r$, where π is the ratio between the circumference and the diameter of a circle. (The value of π is approximately 3.14.) Consider the cases where the tape is wrapped once, twice, etc., around the spool.

CASE 1. $k = 1$

For $k = 1$, the tape is wrapped around the right spool once. So the tape has length equal to the circumference of the spool. Therefore

$$L = 2\pi r,$$

where r is the radius of the spool.

CASE 2. $k = 2$

For $k = 2$, the tape is wrapped around the right spool twice. The length of the first wrap is, as above, $2\pi r$. The radius of the circle for the second wrap is increased by the thickness of the tape, d. See the Figure 2.

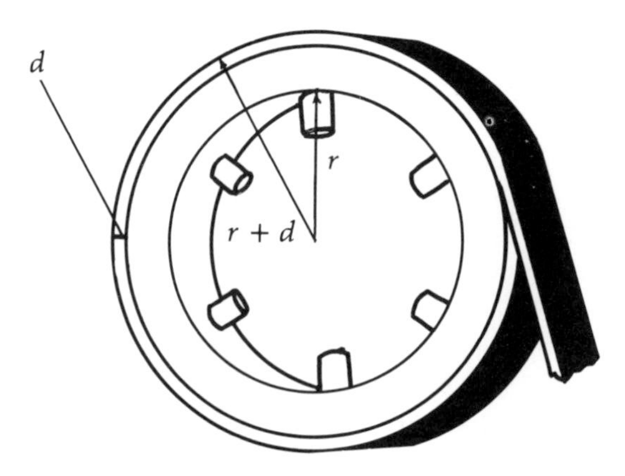

Figure 2.

The length of the second wrap is $2\pi(r+d)$. Therefore,

$$L = 2\pi r + 2\pi(r+d).$$

THE CASE $k = 3$

Obviously, L is the sum of the lengths of three wrappings, the first of radius r, the second of radius $r + d$, and the third of radius $r + 2d$. Therefore,

$$L = 2\pi r + 2\pi(r+d) + 2\pi(r+2d).$$

For k turns around the right spool, we have

$$L = 2\pi r + 2\pi(r+d) + 2\pi(r+2d) + \ldots + 2\pi(r+k-1)d).$$

Factoring out 2π we obtain

$$L = 2\pi(r+(r+d) + (r+2d) + \ldots + (r+(k-1)d)).$$

So

$$L = 2\pi(kr+d(1 + 2 + \ldots + (k-1))). \quad (3)$$

This expression can be further simplified. Let

$$S = 1 + 2 + \ldots + (k-2) + (k-1).$$

Then

$$S = (k-1) + (k-2) + \ldots + 2 + 1.$$

Now add both sides of these equations to obtain

$$2S = k + k + \ldots + k + k.$$

Since the last k appears $k - 1$ times in the right-hand side of this last equation, we have

$$2S = k(k-1)$$

or

$$S = k(k-1)/2.$$

this last equation can be used to simplify (3) to obtain

$$L = 2\pi(kr + k(k-1)d/2)$$

$$= \pi dk^2 + \pi(2r-d)k. \qquad (4)$$

Equation (1) allows us to substitute cn for k in (4) to obtain

$$L = \pi dc^2n^2 + \pi(2r-d)cn.$$

By equation (2), then,

$$T = L/105$$

$$= (\pi dc^2/105)n^2 + (\pi(2r-d)/105)\ cn. \qquad (5)$$

Although equation (5) takes into account the length of each wrapping of the tape around the spool, it ignores that additional length of tape due to the movement of the tape away from the center of the spool. For this reason the equation is not exact. However, it is a good approximation, and this defect is of little concern. A more serious problem with (5) is the difficulty in getting accurate measurements of r and d. The value of d, the tape thickness, is especially difficult to measure. To get a good measurement of r, the radius of the right spool, you might have to disassemble the cassette. Actually, it is impossible to make very accurate measurements of these lengths with a ruler.

The moral of the story is that it is not always good enough to solve a problem by deriving a formula. There must also be a practical way to use the formula. We shall find one in the next section.

Exercise 1;
Use equation (4) to solve the following problem:

A long strip of paper is wrapped around a cardboard cylinder to form a roll of paper whose outer diameter is 10 inches. The diameter of the cylinder is 2 inches. The paper is wrapped around the cylinder 400 times. How long is the paper?

3. Practical Computations

3.1 Predicting Recording Time

Look at the formula (5) derived in Section 2:

$$T = (\pi dc^2/105)n^2 + (\pi(2r-d)/105)\,cn. \tag{5}$$

If we set

$$A = \pi dc^2/105 \tag{6}$$

and

$$B = \pi(2r-d)c/105, \tag{7}$$

then we get

$$T = An^2 + Bn. \tag{8}$$

The accuracy of equation (8) depends upon the accuracy of the evaluations of A and B. If we cannot measure the values of r and d accurately, then we cannot hope to determine A and B accurately from equations (6) and (7).

Our approach is to obtain two sets of values for n and T directly. These give us a system of two equations in the unknowns A and B. Solving this system yields values for A and B. In taking this approach we are substituting the accuracy of a clock with a second hand for the accuracy of a ruler. Since an electric clock with a second hand keeps nearly perfect time, we can get excellent results.

Consider the following example. Suppose after resetting the counter to zero and starting a tape at its beginning, the counter reads 262 at the end of 10 minutes and 640 at the end of 30 minutes. (These figures come from Table 1.) This yields two sets of readings, (n_1, T_1) and (n_2, T_2), where

$$n_1 = 262,\ T_1 = 10,\ n_2 = 640, \text{ and } T_2 = 30.$$

By equation (8),

$$An_1^2 + Bn_1 = T_1$$

$$An_2^2 + Bn_2 = T_2.$$

As a convenience in solving this system, divide the first equation by n_1 and the second by n_2 to obtain

$$An_1 + B = T_1/n_1$$

$$An_2 + B = T_2/n_2.$$

Subtracting the second equation from the first gives

$$A(n_2 - n_1) = T_2/n_2 - T_1/n_1.$$

Thus

$$A = \frac{T_2/n_2 - T_1/n_1}{n_2 - n_1} \tag{9}$$

Substituting the value for A into one of the equations of the system, say the second, we get

$$B = T_2/n_2 - An_2. \tag{10}$$

Since the counter reading is usually three digits, we shall round off all our answers to three significant digits. So in our example,

$$A = \frac{30/640 - 10/262}{640 - 262}$$

$$= 0.0000230$$

and

$$B = 30/640 - (.0000230346)640$$

$$= 0.0321.$$

Thus we have the formula

$$T = 0.0000230\,n^2 + 0.0321\,n. \tag{11}$$

If you use a calculator to evaluate T, it is easier to factor out an n and put equation (11) in the form

$$T = n(0.0000230\,n + 0.0321). \tag{12}$$

For example, to find out how much time has passed when the counter has value 466, we substitute $n = 466$ in equation (12) to obtain

$$T = 466((0.0000230)\,466 + 0.0321)$$

$$= 20.0 \text{ minutes},$$

in agreement with Table 1.

Exercise 2.
Assume the values of A and B in the example of this section. Suppose you reset the counter to zero and start recording at the beginning of a cassette with 45 minutes on each side. When the counter reads 500, how much more recording time do you have left on the side?

Exercise 3.
After resetting the counter to zero and recording from the beginning of a new tape you find that the counter reads 151 at the end of 5 minutes and 392 at the end of 15 minutes. How much time have you spent recording when the counter reads 500? 600?

Exercise 4.
Substitute $A = 0.0000230$, $B = 0.0321$, and $c = 4/3$ into equations (6) and (7) and solve the resulting equations for d, the tape thickness, and r, the radius of the right spool.

3.2 The Inverse Relation

The equation

$$T = An^2 + Bn \tag{8}$$

allows us to determine the value for T given a value for n. It is also useful to be able to solve the opposite problem:

> Given a value for T, the time that the tape has been running, find the value for n, the counter reading.

Here we must regard equation (8) as a quadratic equation in unknown n. Recall that the solutions to a quadratic equation

$$ax^2 + bx + c = 0$$

in unknown x are given by the quadratic formula

$$x = \frac{-b \pm \sqrt{b^2 - 4ac}}{2a} \tag{13}$$

If we rewrite equation (8) in quadratic form

$$An^2 + Bn - T = 0,$$

we can apply equation (13) to solve for n. Since n must be positive, we can reject the negative solution to obtain

$$n = \frac{-B + \sqrt{B^2 + 4AT}}{2A} \tag{14}$$

If $A = 0.0000230$ and $B = 0.0321$, what will the counter read after the tape has run for 4 minutes? Since $T = 4$,

$$n = \frac{-0.0321 + \sqrt{(0.0321)^2 + 4(0.0000230)4}}{2(0.0000230)}$$

$$= 115.$$

Exercise 5.
Suppose that after resetting the counter to zero and recording from the beginning on a new tape, the counter reads 145 at the end of 5 minutes and 270 at the end of 10 minutes. What will the counter read at the end of 20 minutes of recording?

Exercise 6.
In this exercise we look at an alternate derivation of a formula relating L and k.

We are given a tape of length L, width w, and thickness d. Unrolled it is a long rectangular solid whose volume is

$$V = Lwd. \tag{15}$$

See Figure 3(a). Now suppose that this tape is wrapped around a spool of radius r and the wrapping requires k turns. See Figure 3(b).

(a) Show that the volume of the tape is given by

$$V = \pi(r + kd)^2 w - \pi r^2 w. \tag{16}$$

Hint: Use the fact that the volume of a right circular cylinder of radius R and height h is $\pi R^2 h$.

(b) Use (15) and (16) and solve for L to obtain

$$L = \pi d k^2 + 2\pi r k.$$

(a)

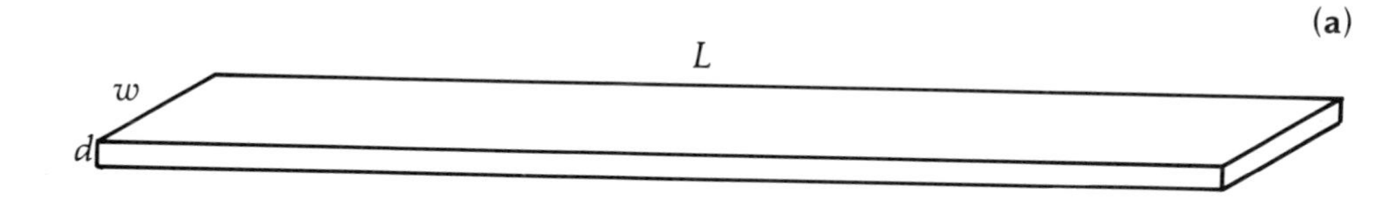

(b)

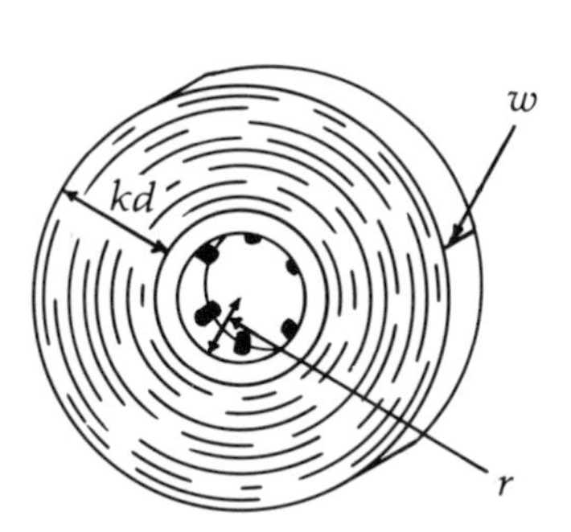

Figure 3. The tape of length L, width w, and thickness d in (a) is wrapped k times around a spool of radius r in (b).

4. Answers to Exercises

1. 7530 inches

2. 23.2 minutes

3. 20.3 minutes, 25.6 minutes

4. d = 0.000432 inches, r = 0.403 inches

5. n = 483

UMAP

Modules in Undergraduate Mathematics and its Applications

Module 651

The Unit of Analysis and the Independence of Observations

Thomas R. Knapp

Published in cooperation with the Society for Industrial and Applied Mathematics, the Mathematical Association of America, the National Council of Teachers of Mathematics, the American Mathematical Association of Two-Year Colleges, and The Institute of Management Sciences.

COMAP

INTERMODULAR DESCRIPTION SHEET:	UMAP Unit 651
TITLE:	The Unit of Analysis and the Independence of Observations
AUTHOR:	Thomas R. Knapp Graduate School of Education and Human Development University of Rochester Rochester, New York 14627
MATH FIELD:	Statistics
APPLICATION FIELD:	Social science
TARGET AUDIENCE:	Students in a statistics course.
ABSTRACT:	This module addresses two problems that have received insufficient attention in the statistical literature, namely, the selection of an appropriate unit of analysis and the determination of the independence of a set of observations that comprise aggregated data. An example from research on twins is used throughout the module. Exercises, an annotated bibliography, and an end-of-module quiz are included.
PREREQUISITES:	1. Knowledge of basic descriptive statistics (mean, variance, etc.) and the standard normal distribution. 2. Ability to calculate means and variances. 3. Exposure to ordinary correlation and regression (one X, one Y). 4. Familiarity with rules for manipulating summation signs.

The UMAP Journal, Vol. V, No. 3, 1984

The Unit of Analysis and the Independence of Observations

Thomas R. Knapp

Graduate School of Education and Human Development
University of Rochester
Rochester, New York 14627

Table of Contents

1. INTRODUCTION 1
2. AN EXAMPLE INVOLVING REAL DATA 2
3. THE UNIT OF ANALYSIS 4
4. THE INDEPENDENCE OF OBSERVATIONS 4
5. SOME ADDITIONAL COMMENTS REGARDING I 8
6. TOTAL, BETWEEN, AND WITHIN CORRELATION COEFFICIENTS 9
7. BUT WHICH CORRELATION IS THE "RIGHT" CORRELATION FOR THE TWIN DATA? 11
8. HOW ABOUT CONTROLLED EXPERIMENTS? 11
9. SUMMARY 12
10. ANNOTATED BIBLIOGRAPHY 12
11. END-OF-MODULE QUIZ 15
12. ANSWERS TO EXERCISES 17
13. ANSWERS TO QUIZ 18
14. APPENDIX 19

Modules and Monographs in Undergraduate Mathematics and its Applications Project (UMAP)

The goal of UMAP was to develop, through a community of users and developers, a system of instructional modules in undergraduate mathematics and its applications to be used to supplement existing courses and from which complete courses may eventually be built.

The Project was guided by a National Advisory Board of mathematicians, scientists, and educators. UMAP was funded by a grant from the National Science Foundation and is now supported by the Consortium for Mathematics and Its Applications, Inc. (COMAP), a nonprofit corporation engaged in research and development in mathematics education.

The project would like to thank David Herr of the University of North Carolina—Greensboro, Robert Rippey of the University of Connecticut Health Center, and Gary Simon of New York University for their reviews, and all others who assisted in the production of this unit.

This material was prepared with the partial support of National Science Foundation Grant No. SPE8304192. Recommendations expressed are those of the authors and do not necessarily reflect the views of the NSF or the copyright holder.

1. Introduction

One of the most common questions in scientific research is "What is the relationship between X and Y?" where X and Y are variables such as anxiety and achievement, height and weight, race and intelligence, etc. There are several sticky issues that must be faced in research that is concerned with the relationships between variables. This module addresses two of them: the unit of analysis and the independence of observations.

Consider the research question "What is the relationship between height and weight?" and the following sets of fictitious data.

Set A	Average Height	Average Weight	Set B	Height	Weight
Jockeys	5'0"	110#	Joe (a jockey)	4'11"	120#
Boxers	5'8"	170#	Sam (a boxer)	5'9"	160#
Football Players	5'11"	200#	Bill (a fb. play.)	5'10"	210#
Basketball Players	6'3"	190#	Jim (a bb. play.)	6'4"	180#

If these were the only two sets of data available, which one would you use? That's a very tough question, since it involves a number of decisions including the choice of a *unit of analysis*, which for this example is either the individual (Set B) or the group (Set A).

Or consider the following sets of data for the same research question (the data are also fictitious):

Set C	Height	Weight	Set D	Height	Weight
Sue	5'4"	125#	Sue	5'4"	125#
Ginny	5'3"	135#	Ginny	5'3"	135#
Sally	5'8"	150#	Ginny	5'3"	135#
Ellen	5'6"	120#	Sally	5'8"	150#

Again assuming these were the only two sets of data available, which one would you use? That question is easier to answer, since in Set C there are four different people for whom you have data whereas in Set D Ginny is counted twice. The observations in Set C appear to be *independent* but the observations in Set D are not, and therefore the nod goes to Set C.

2. An Example Involving Real Data

Now suppose that you were interested in the correlation between height and weight for *twins*. To be more specific, let's say that you wanted to determine the direction and the magnitude of the relationship between height and weight for seven pairs of 16-year-old, black, female, identical twins. (See Table 1 for the data, which were actually part of a research study.) How would you proceed?

The best way to start is to plot the data. On the usual set of coordinate axes with X = height and Y = weight, you would plot $(X, Y) = (68{,}148)$ for the first person, A, in the study. So far, so good. But when you go to plot the data for A's twin, a, for whom $X = 67$ and $Y = 137$, it occurs to you that you should have some way of showing that these two data points represent people who are twins of one another, in order to distinguish them from members of other twin-pairs. But how? One way would be to plot both points with the same colored pen or pencil, e.g., red, and to use a different color for each twin-pair. Alternatively, or additionally, you could carefully label the points A and a, B and b, etc. The latter approach will be adopted here; the resulting scatter diagram is shown in Figure 1. (Note that points B and e are the same point. Real data are like that!)

Table 1

The Heights and Weights of Seven Pairs of 16-year-old, Black, Female, Identical Twins*

Pair	X Heights (in inches)		Y Weights (in pounds)	
1 (Aa)	A:68	a:67	A:148	a:137
2 (Bb)	B:65	b:67	B:124	b:126
3 (Cc)	C:63	c:63	C:118	c:126
4 (Dd)	D:66	d:64	D:131	d:120
5 (Ee)	E:66	e:65	E:123	e:124
6 (Ff)	F:62	f:63	F:119	f:130
7 (Gg)	G:66	g:66	G:114	g:104

*Source: Osborne, R. T. *Twins: Black and White*, pages 195-200.

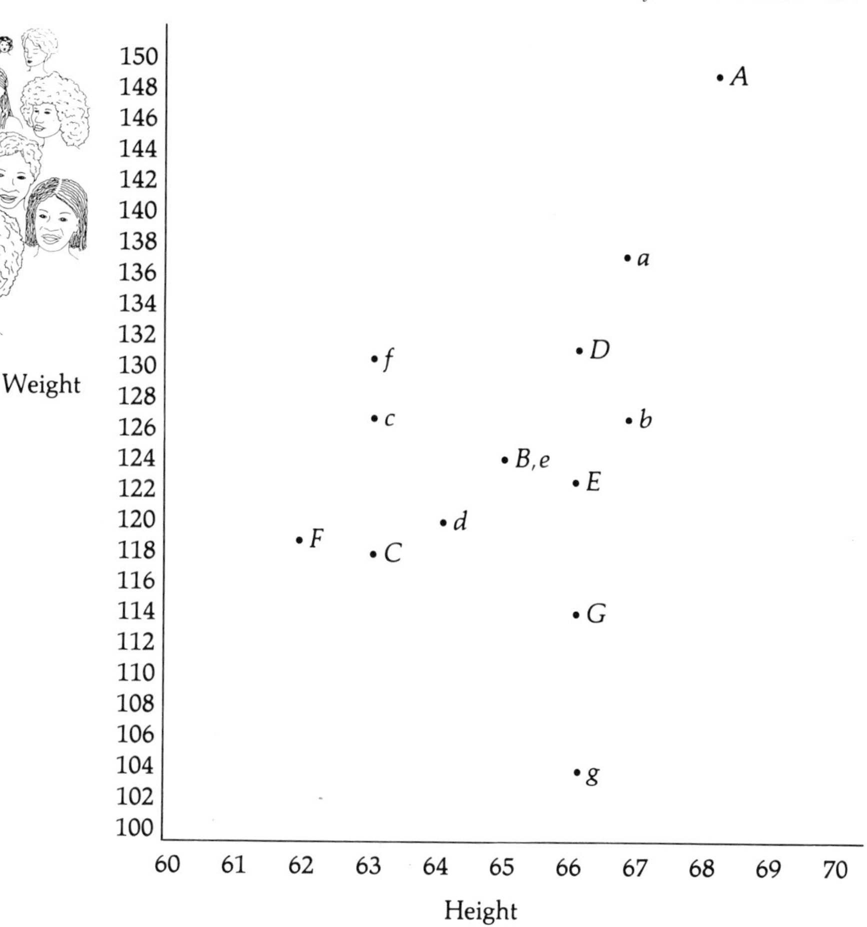

Figure 1. Scatter Diagram for the Twin Data

Now what? The swarm of points looks reasonably elliptical and let's say that you were interested in a linear relationship all along, so you decide that the Pearson product-moment correlation coefficient is what you want to calculate. You look up the formula in a statistics book and you pick out

$$\frac{\Sigma xy}{N\sigma_x \sigma_y}$$

(There are other expressions which are mathematically equivalent to this one.) That looks pretty straightforward: x is the deviation of each height from the mean of the heights; y is the deviation of each weight from the mean of the weights, σ_x and σ_y are the standard deviations of the heights and the weights, respectively; and N is . . . , N is . . . , *hmm*, what is N? The number of *twins* or the number of *twin-pairs*? 14 or 7? You have 14 data points but something tells you that is is not like having paired values of heights and weights for 14 mutually unrelated people.

So you go back to the drawing board. Perhaps you should average the heights and the weights for each twin-pair and plot those seven points. No, that would ignore some potentially interesting between-person-within-pair variability. How about choosing one twin at random from each pair (by tossing a coin, for example) and plotting *those* points? No, that would also throw away data. Perhaps you should plot the weight of one member of each pair against the height of the other member. No, that sounds silly.

What started out as an ostensibly simple problem has become a nightmare. Why? Because the determination of the correlation between height and weight for twins raises some very subtle problems regarding the unit of analysis *and* the independence of observations. These matters have not been given much attention in the statistical literature and, perhaps consequently, have been badly botched in many research studies.

3. The Unit of Analysis

Of crucial importance for any research investigation is the specification of the "thing" on which measurements have been obtained and that will serve as the unit of analysis for the data. For the heights-and-weights-of-twins example, is the unit of principal concern the person or the pair of persons? Does it matter? You bet it does, as W.S. Robinson pointed out in 1950 in his classic article in the *American Sociological Review*, "Ecological correlations and the behavior of individuals," (sociologists call correlations for which something other than an individual person is the unit of analysis "ecological"), and as the formulas in Section 6 of this module will attest. But what if the data are not available for the desired unit of analysis, say the student, and are available for some other unit, say the classroom? Can you estimate the correlation you want? Maybe, but more about that later.

4. The Independence of Observations

The phrase "the observations must be independent" comes up a lot in the study of statistics, but with very few exceptions the student is never really told what this means, particularly in the real world. How do you know whether or not two or more observations are independent? What do you do if they're not?

In one of the few statistics textbooks that includes a discussion of independence of observations, Helen Walker and Joseph Lev state: "Two observations are considered to be independent when information about one of them provides no clue whatever as to the other." (Walker, H. M. and Lev, J. *Statistical Inference*, page 14.) They go on to give a few examples of independent and non-independent (i.e., dependent) observations, including one example of families in Iowa where observations are independent with respect to one population but are dependent with respect to another. But even they leave the student hanging regarding why there is independence in one case and not in the other.

In a paper presented at the annual meeting of the American Educational Research Association in 1976, Linda Glendening provided the basis for an actual measure of independence. It is well known that for an infinite population, if X is distributed with mean μ and variance σ^2 then

$$\overline{X} = \frac{1}{n} \sum_{i=1}^{n} X_i$$

is distributed with mean μ and variance σ^2/n for random samples of n independent observations. That is, the sample means have the same mean as the original observations but a variance that is one n^{th} the size of the variance of the original observations. Turning this result inside out, so to speak, Glendening defined statistical independence as follows (in my terminology and notation, not hers): A set of observations upon which some sort of aggregation has been imposed comprises a set of independent observations if the variance of the aggregate means is equal to the within-aggregate variance divided by the number of observations per aggregate, i.e., $\sigma_m^2 = \sigma_w^2/n$, for $n \geq 2$, $\sigma_m^2 \neq 0$, $\sigma_w^2 \neq 0$. This definition suggests the following measure of independence:

$$I = \frac{\sigma_m^2}{\dfrac{\sigma_w^2}{n}} = \frac{n\sigma_m^2}{\sigma_w^2} \quad . \qquad (4.1)$$

When $I = 1$ the observations are "perfectly" independent; when $I > 1$ there is greater variability among the aggregate means than there would be for independent observations; when $I > 1$ there is less variability among the aggregate means than there would be for independent observations.

Table 2 contains some simple examples of four sets of fictitious data, the first three of which consist of dependent observations (I is greater than one for set #1 and set #3, and is less than one for set #2) and the last of which consists of independent observations.

Table 2

Examples of Dependent and Independent Observations
(Two aggregates; n = 3 observations per aggregate)

Data set	Aggregate #1	Aggregate #2	σ_m^2	σ_w^2	$\frac{\sigma_w^2}{n}$	I
1	1,2,3	4,5,6	$\frac{9}{4}$	$\frac{2}{3}$	$\frac{2}{9}$	$\frac{81}{8}$
2	1,3,5	2,4,6	$\frac{1}{4}$	$\frac{8}{3}$	$\frac{8}{9}$	$\frac{9}{32}$
3	1,2,3	2,3,4	$\frac{1}{4}$	$\frac{2}{3}$	$\frac{2}{9}$	$\frac{9}{8}$
4	1,7,13	5,11,17	4	12	4	1

For the twin data (see above) there are seven aggregates and two observations per aggregate. (Each twin-pair is an aggregate.) With respect to the height variable, $\sigma_m^2 = 2.67$. That is, the variance of the seven mean heights 67.5, 66.0, 63.0, 65.0, 65.5, 62.5, and 66.0 is 2.67 (see Table 3). The within-aggregate variances differ from aggregate to aggregate (ranging from 0.00 to 1.00), but the average (mean) of these variances is 0.393. Letting $\sigma_w^2 = 0.393$ and with $n = 2$, we have

$$I = (2)(2.67)/0.393 = 13.6.$$

Since I is considerably greater than 1 for these data, the observations are far from independent of one another.

As further evidence that

$$I = \frac{n\sigma_m^2}{\sigma_w^2}$$

is a defensible measure of statistical independence, consider the data in Table 4, which are the first 50 random normal deviates in the 1955 Rand tables arranged arbitrarily in five aggregates of ten observations each. Here the population variance σ^2 is known to be 1 since the observations were sampled from the standard normal distribution, but the within-aggregate variances vary, by chance, from 0.440 to 1.913 with a mean of 0.973. The variance of the five aggregate means, σ_m^2, is equal to 0.100 and I is therefore (10) (0.100)/0.973 = 1.028. This is not exactly equal to 1, even for this "obviously independent" situation (again because of chance), but it is a heck of a lot closer to 1 than the value of I for the heights of the twins.

Table 3

Independence Calculations for the Twin Data

(twin-pair) Aggregate	Heights	Weights	Mean Heights	Within-pair Height Variances	Mean Weights	Within-pair Weight Variances
1(Aa)	68 and 67	148 and 137	67.5	.25	142.5	30.25
2(Bb)	64 and 67	124 and 126	66.0	1.00	125.0	1.00
3(Cc)	63 and 63	118 and 126	63.0	.00	122.0	16.00
4(Dd)	66 and 64	131 and 120	65.0	1.00	125.5	30.25
5(Ee)	66 and 65	123 and 124	65.5	.25	123.5	.25
6(Ff)	62 and 63	119 and 130	62.5	.25	124.5	30.25
7(Gg)	66 and 66	114 and 104	66.0	.00	109.0	25.00

For height: $\sigma_m^2 = 2.67$

$\sigma_w^2 = 0.393 \qquad n = 2$

$I = (2)(2.67)/0.393 = 13.6$

For weight: $\sigma_m^2 = 81.8$

$\sigma_w^2 = 19.0 \qquad n = 2$

$I = (2)(81.8)/19.0 = 8.61$

Exercises

1. For the weights of the seven twin-pairs, $I = 8.61$, which is closer to 1 than the I of 13.6 for the heights? Does that make sense? Why or why not?

2. If the data in Table 4 are transposed and considered as ten aggregates of five observations each, it turns out that $\sigma_m^2 = 0.045$, $\sigma_w^2 = 1.029$, and $I = 0.218$. Does it make sense that this I differs from 1 by more than the I for the five aggregates, ten observations data does? Why or why not?

Table 4

Fifty Random Normal Deviates*

Aggregate	Observations	Mean	Variance
1	−1.276, −1.218, −.453, −.350, −.723 .676, −1.099, −.314, −.394, −.633	−.434	.440
2	−.318, −.799, −1.664, 1.391, .382, .733 .653, .219, −.681, 1.129	.105	.823
3	−1.377, −1.257, .495, −.139, −.854, .428, −1.322, −.315, −.732, −1.348	−.642	.473
4	2.334, −.337, −1.955, −.636, −1.318, −.433, .545, .428, −.297, .276	−.139	1.216
5	−1.136, .642, 3.436, −1.667, .847, −1.173, −.355, .035, .359, .930	.192	1.913

$I = (10)(0.100)/0.973$
$= 1.028$

$\sigma_m^2 = .100$ $\sigma_w^2 = .973$
$n = 10$

*Source: The Rand Corporation. *A Million Random Digits with 100,000 Normal Deviates,* page 1 of the second set of tables.

5. Some Additional Comments Regarding *I*

The within-aggregate variance σ_w^2 was constant across aggregates for the four fictitious data examples in Table 2 but varied from aggregate to aggregate for the data in Table 3 and Table 4. The latter will always be the case for real data, and it seems natural in the formula for I to let σ_w^2 be equal to the mean of the several within-aggregate variances, as we have done. One might want to first ask whether or not the within-aggregate variances are "averageable" before computing I. Although there are several homogeneity-of variance tests for sample data (none of which are terribly sensitive, however), when you have complete population data for aggregates and observations within aggregates, the "averageability" of the within-aggregate variances must be a judgment call. (The within-pair variances of the weights of the seven twin-pairs range from a very small 0.25 to a very large 30.25 and strongly suggest "non-averageability," but with only two observations per aggregate almost anything can happen.)

For most real data n also varies from aggregate to aggregate. (For twin research n is necessarily constant since there must be two persons per aggregate.) A reasonable thing to do in such an eventuality would be to calculate σ^2_{wj}/n_j for each aggregate j, average *those*, and use that for σ^2_w/n in calculating I. And in order to be consistent, σ^2_m in the numerator of I should be the *weighted* variance of the means of the aggregates, obtained by calculating

$$\frac{\Sigma n_j (\mu_j - \mu_t)^2}{\Sigma n_j}$$

where μ_j is the mean for aggregate j and μ_t is the grand (total) mean of all the observations.

6. Total, Between, and Within Correlation Coefficients

According to Robinson, the correlation coefficients for two variables across-aggregates (the "total" correlation), between-aggregates, and within-aggregates are tied together like this:

$$\rho_t = \eta_x \eta_y \rho_b + \sqrt{1 - \eta_x^2}\,\sqrt{1 - \eta_y^2}\;\rho_w. \tag{6.1}$$

Here ρ_t is the total correlation with the individual "object" as the unit of analysis, ρ_b is the between-aggregate correlation with the aggregate itself as the unit of analysis, ρ_w is the "pooled" within-aggregate correlation with the individual object as the unit of analysis, and η_x^2 and η_y^2 are the "correlation ratios" (also called "intraclass correlations") for X and Y, respectively. Computationally,

$$\rho_t = \frac{\Sigma x_t y_t}{\sqrt{\Sigma x_t^2}\sqrt{\Sigma y_t^2}} \qquad (6.2)$$

$$\rho_b = \frac{\Sigma n_j x_{b_j} y_{b_j}}{\sqrt{\Sigma n_j x_{b_j}^2}\sqrt{\Sigma n_j y_{b_j}^2}} \qquad (6.3)$$

$$\rho_w = \frac{\Sigma x_w y_w}{\sqrt{\Sigma x_w^2}\sqrt{\Sigma y_w^2}} \qquad (6.4)$$

$$\eta_x^2 = \frac{\Sigma n_j x_{b_j}^2}{\Sigma x_t^2} \qquad (6.5)$$

$$\eta_y^2 = \frac{\Sigma n_j y_{b_j}^2}{\Sigma y_t^2} \qquad (6.6)$$

The x's and y's are deviations of the individual observations from the grand mean (total), the deviations of the aggregate means from the grand mean (between), and the deviations of the individual observations from their respective means (within). The η^2's are measures of the amount of between-aggregate separation.

For the twin data we have:

$\rho_t = .375$

$\rho_b = .333$

$\rho_w = .614$

$\eta_x^2 = .872$

$\eta_y^2 = .812$

Suppose you know only ρ_b for a given problem? Could you estimate ρ_w, for example? Only if you had a pretty good idea regarding how the aggregates spread apart on X and Y (i.e., you had some good guesses for η_x^2 and η_y^2) *and* you had an equally good feel for ρ_t. Solving for ρ_w in Eq. 1 and plugging in the value of ρ_b and the approximations to η_x^2, η_y^2, and ρ_t would do the job.

Exercise

3. Calculate ρ_b for the "Set A" data on page 1. (Assume that the n_j are all the same [equal to n] for the four groups of athletes. What happens to the formula for ρ_b if that is the case?)

Estimate η_x^2 and η_y^2, and use the correlation for the "Set B" data as an estimate of ρ_t. Then use Eq. (6.1) to estimate ρ_w.

7. But Which Correlation is the "Right" Correlation for the Twin Data?

As you may have already anticipated, there is no such thing as the "right" correlation. The answer to the question "What is the correlation between height and weight for our seven pairs of 16-year-old black, female, identical twins?" is 0.375 if the twinship of the various pairs is ignored; it is 0.333 if it makes substantive sense to consider the twin-pair as the unit of analysis; and the correlation "averages out" to 0.614 within twin-pairs. (All seven of the respective within-aggregate correlations must be $+1$, -1, or indeterminate, since $n = 2$.)

The following argument makes a strong case for ρ_w, however. The fact that the data are available by aggregates (twin-pairs) is actually annoying, and it would be nice to be able to control for that. There is a technique, called *partial correlation*, which permits answers to research questions such as "What is the relationship between X and Y with Z statistically controlled (held constant, partialled out, taken into account, etc.)?" The technique can be generalized to more than one control variable, i.e., it is possible to control for Z_1, Z_2, Z_3, etc. For the twin data, a re-phrasing of the research question would be, "What is the relationship between height and weight, with twin-pair designation partialled out?" It turns out that the partial correlation between X and Y (X = height and Y = weight) with Z_1, Z_2, Z_3, Z_4, Z_5, and Z_6 partialled out (it takes six "dummy variables" to "code" the twin-pair information for this example) is mathematically identical to the ρ_w for these data, i.e., 0.614. (See the Appendix.)

8. How About Controlled Experiments?

Data aggregation can come about in a wide variety of ways. In a controlled experiment, for example, after subjects are randomly assigned to treatment conditions, they are often further allocated to

groups (classrooms, say) for convenience in implementing the treatments. (Subjects are said to be "nested" within groups.) Interest lies in the research question "What is the relationship between type of treatment and some dependent variable?" and once again considerations involving the unit of analysis (subject within group? subject across groups? groups?) and the independence of observations (are the within-treatment observations reasonably independent?) raise their ugly heads. Just as in non-experimental research, the dilemma can be resolved in favor of ρ_w since the groups within treatments are methodological nuisances and the research question of principal interest is probably the rephrased "What is the relationship between type of treatment and some dependent variable, with within-treatment grouping partialled out?"

9. Summary

In this module I have tried to call your attention to the unit-of-analysis problem in scientific research and to clarify what is meant by independent observations. I have introduced a measure of the degree of independence of a set of observations for which there is some sort of aggregation. The formula that shows how between-aggregate, within-aggregate, and total correlation coefficients are related to one another was presented near the end of the module. The within-aggregate correlation coefficient was shown to be equal to the partial correlation coefficient controlling for aggregate. An example involving twin-pair data was used throughout the module for illustrative purposes.

The next time you talk about the correlation between two variables, ask yourself what the unit of analysis is and whether or not the observations are independent. These questions are guaranteed to make you feel uncomfortable, but you will have helped one person (yourself) to think more carefully about research methodology. And that's a start.

10. Annotated Bibliography

Burstein, L. "The analysis of multilevel data in educational research and evaluation." Ch. 4 in Berliner, D.C. (ed.) *Review of Research in Education*, Vol. 8, Itasca, IL: Peacock, 1980.

Cronbach, L. J. *Research on Classrooms and Schools: Formulation of Questions, Design and Analysis.* Occasional paper of the Stanford Evaluation Consortium, Stanford University, 1976.

These two recent monographs deal primarily with unit-of-analysis problems in research on the effectiveness of schooling.

Glass, G. V and Stanley, J. C. *Statistical Methods in Education and Psychology.* Englewood Cliffs, NJ: Prentice-Hall, 1970.

On pages 501-508, these authors give an excellent discussion of the within-treatment dependence of observations in many educational experiments.

Glendening, L. "The effects of correlated units of analysis: choosing the appropriate unit." Paper presented at the annual meeting of the American Educational Research Association, San Francisco, April 1976.

This is the paper that was cited in this module as the source of the quantitative definition of statistical independence.

Knapp, T. R. "The unit-of-analysis problem in applications of simple correlation analysis to educational research." *Journal of Educational Statistics,* 1977, *2*, 171-186.

As the title of this article suggests, applications of ordinary correlation analysis involving different units of analysis are discussed, including those involving two levels of aggregation (for example, students "nested" within classrooms and classrooms "nested" within schools).

Osborne, R. T. *Twins: Black and White.* Athens, GA: Foundation for Human Understanding, 1980.

This book is the source of the twin data used throughout this module.

Pedhazur, E. J. *Multiple Regression in Behavioral Research* (second edition). New York: Holt, Rinehart, and Winston, 1982.

This popular textbook for regression analysis contains a fine section on the unit of analysis (pp. 526-547). The author ties the unit-of-analysis problem to the analysis of covariance, rather than to partial correlation analysis as in this module. But those two analyses really address the same question, viz. the relationship between two variables that is "over and above" what could be predicted from the variable(s) that one wishes to control.

Rand Corporation (The). *A Million Random Digits with 100,000 Normal Deviates.* Glencoe, IL: The Free Press, 1955.

A book with an extensive cast of characters but not much of a plot, it is the source of the 50 random normal deviates in Table 4 of this model.

Robinson, W. S. "Ecological correlations and the behavior of individuals." *American Sociological Review,* 1950, *15,* 351-357.

This article is known by almost all sociologists as the first paper to call to the attention of researchers in that discipline the error of using between-aggregate correlation coefficients as substitutes for correlation coefficients for individuals.

Sirotnik, K. A. "Psychometric implications of the unit-of-analysis problem (with examples from the measurement of organizational climate)." *Journal of Educational Measurement,* 1980, *17,* 245-282.

Sirotnik treats in considerable detail some of the unit-of-analysis problems that arise in the measurement aspects of a research study.

Walker, H. M. and Lev, J. *Statistical Inference.* New York: Holt (now Holt, Rinehart, and Winston), 1953.

An exception to the rule among statistics textbooks, this book contains a good discussion (pp. 14-16) of the independence of observations.

Yule, G. U. and Kendall, M. G. *An Introduction to the Theory of Statistics,* 14th edition. New York: Hafner, 1968.

This durable textbook treats the unit-of-analysis problem in the context of combinable geographical areas for determining the correlation between variables such as wheat-yield and potato-yield.

11. End-of-Module Quiz

1. Here are the heights (in inches) of 23 students I recently taught in a statistics course:

Males	Females
75	66
73	67
71	62
65	65
72	62
68	64
67	66
67	68
	62
	68
	63
	66
	64
	67
	67

   a. Before carrying out any calculations, would you expect these observations to be independent or not? Why?

   b. Compute I for these data. (What did you use for σ_m^2 and σ_w^2/n?)

   c. Are those observations statistically independent, is there more between-aggregate variation than there would be for independent observations, or is there less between-aggregate variation than there would be for independent observations? How does this jibe with your answer to Part a?

2. Check the claims made in Exercise 2 that the transposed random normal deviates have $\sigma_m^2 = 0.045$, $\sigma_w^2 = 1.029$, and $I = 0.218$.

3. Consider the following set of fictitious data (Pedhazur, E. J. *Multiple Regression in Behavioral Research,* page 543):

Group	X	Y	Group	X	Y
1	1	6	3	7	20
	1	9		7	22
	2	8		9	24
	3	8		9	26
	3	12		10	24
	4	12		11	25
	4	10		11	28
	5	8		12	27
	5	12		13	29
	6	13		13	26
2	4	13	**4**	7	27
	4	16		8	28
	5	15		8	25
	6	16		9	27
	6	19		9	31
	8	17		10	29
	8	19		10	32
	9	23		12	30
	10	19		12	32
	10	22		14	33

a. If X were aspiration and Y were achievement (Pedhazur's first substantive situation), what unit of analysis do you think would be most appropriate when determining the relationship between X and Y, if the four groups were four classrooms—person within aggregate, person across aggregates, or aggregate? Why?

b. If X were anxiety and Y were productivity (Pedhazur's second substantive situation), and the groups were factory teams, would your choice be the same?

c. If you wanted to calculate ρ_w for these data (this does not necessarily mean that person within-aggregate is the "right" answer to Parts a and b of this question) and you preferred to use the partial correlation formula, how many "dummy" variables would you need to "code" the aggregates (groups)? Why?

d. Are η_x^2 and η_y^2 small or large for these data? (Answer this question without using any formulas to calculate them.)

e. Which would you guess would be larger, ρ_t or ρ_w? (Again, answer this question without actually carrying out any formal calculations.) Why?

4. (Bonus question) For the twin data, we calculated I separately for the heights and weights. Can you think of a way to define I for bivariate data so that you get *one* measure for both variables considered simultaneously?

12. Answers to Exercises

1. Yes. Since weight is more subject to environmental influences than height is, the similarity of the weights of the seven twin-pairs to the weights of fourteen unrelated individuals should be greater than the similarity of their heights.

2. Yes. There is more "wiggle" in ten aggregates of five observations each than there is in five aggregates of ten observations each, i.e.,chance can play a greater role regarding what you get vis-á-vis what you expect to get because of the smaller *within-aggregate* size. (The number of aggregates is essentially irrelevant here.)

3. a. I get $\rho_b = 0.931$, which is not surprising (only one "reversal" in the means). If all $n_j = n$, then

$$\rho_b = \frac{\Sigma x_{b_j} y_{b_j}}{\sqrt{\Sigma x_{b_j}^2}\sqrt{\Sigma y_{b_j}^2}}$$

b. (One of many possible answers) The numbers η_x^2 and η_y^2 should both be pretty big, say about 0.4, since there should be sizable between-aggregate variation relative to the total variation. The value of ρ_t is 0.707 for the "Set B" data, which is not surprising either, since there is only one "reversal" in the rank-ordering of the people on the two variables. Plugging these into Eq. 1 gives $\rho_w = 0.558$.

13. Answers to Quiz

1. a. Not independent. Each male height should resemble other male heights more than it should resemble any of the female heights, since males are in general taller than females.

   b. I get $I = 5.67$. (I used the weighted formula for σ_m^2 and got 4.84. I averaged $\sigma_1^2/n_1 = 11.32/8$ and $\sigma_2^2/n_2 = 4.38/15$ and got 0.854.)

   c. There is *more* between-aggregate variation than there would be for independent observations.

2. Yep, that's what they are, all right. (Don't worry if you're off by one or two in the least significant digits.) There sure is an imbalance of negative numbers in that set of 50 random normal deviates, isn't there?

3. a. Either person within aggregate or person across aggregates, since individual persons (students) aspire and achieve; aggregates (classrooms) don't.

   b. Here, a stronger case can be made for the aggregate (team) as the appropriate unit of analysis, particularly if the team members work closely together to turn out a finished product. But the individual person (factory worker), preferably within team, is also a reasonable choice. Some methodologists, e.g., Burstein and Cronbach (see references above) suggest carrying out both analyses and interpreting the findings accordingly.

   c. Three (one less than the number of groups). The first group will have "scores" of 1, 0, and 0 on Z_1, Z_2, and Z_3, respectively. The second group will have "scores" of 0, 1, and 0; the third group will have "scores" of 0, 0, and 1; and the fourth group will have "scores" of 0, 0, and 0.

   d. Very large. Group 1 has very low scores on both X and Y, Groups 2 and 3 have intermediate scores on both variables, and Group 4 has very high scores, so the groups are spread quite far apart. (η_x^2 is actually 0.650 and η_y^2 is 0.893.)

e. ρ_t. There are a few "reversals" in rank-order within each group and there is a generally steady increase in both X and Y from the top to the bottom of the list, i.e., from the first person in Group 1 through the last person in Group 4. (ρ_t is actually 0.890 and ρ_w is 0.788.)

4. The bivariate analogue of a variance is the determinant of the variance-covariance matrix (if those terms mean anything to you). So how about something like the following?

$$I = \begin{vmatrix} n \times \dfrac{\sigma_{m_x}^2}{\sigma_{w_x}^2} & n \times \dfrac{\text{cov}(m_x, m_y)}{\text{cov}(x,y)} \\ n \times \dfrac{\text{cov}(m_x, m_y)}{\text{cov}(x,y)} & n \times \dfrac{\sigma_{m_y}^2}{\sigma_{w_y}^2} \end{vmatrix}$$

Here, the upper left-hand corner element and the lower right-hand corner element are the I's for the two variables separately, while the off-diagonal expression for both the upper right-hand element and the lower left-hand element is the covariance of the x and y means for the k aggregates divided by one n^{th} of the "pooled" within-aggregate covariance. This formula is appropriate only when n and the σ_w^2's are constant across aggregates and would have to be altered accordingly for variable n and/or σ_w^2.

14. Appendix

Proof that $\rho_{xy.z} = \rho_w$ for one grouping variable and two equal-sized groups.

Although it can be shown in general that the "pooled' within-aggregate correlation is always equal to the partial correlation when controlling for *any number* of variables that define the aggregation, the proof given here is for the simplest case of one dichotomous grouping variable and equal group sizes. A "score" of 1 on the grouping variable, Z, will be used to designate membership in the first group (aggregate) and a "score" of 0 will be used to designate membership in the second group (aggregate). For this special case the data can be displayed as follows:

Individual	Z	X	Y
1	1	↓	↓
2	1		
⋮	⋮		
$\frac{N}{2}$	1		
$\frac{N}{2}+1$	0		
$\frac{N}{2}+2$	0		
⋮	⋮		
N	0		

The formula for the partial correlation coefficient between X and Y, controlling for Z, is

$$\rho_{xy.z} = \frac{\rho_{xy} - \rho_{xz}\,\rho_{yz}}{\sqrt{1-\rho^2_{xz}}\ \sqrt{1-\rho^2_{yz}}}$$

But

$$\rho_{xy} = \rho_t = \eta_x \eta_y\, \rho_b + \sqrt{1-\eta_x^2}\ \sqrt{1-\eta_y^2}\quad \rho_w$$

by formula (6.1). Furthermore, since the standard deviation of any uniform dichotomy is 1/2,

$$\rho_{xz} = \frac{\Sigma xz}{N\sigma_x\sigma_z} = \frac{\Sigma xz}{N\sigma_x\ (1/2)}$$

$$= \frac{2}{N\sigma_x}\ \Sigma xz = \frac{2}{N\sigma_x}\ \Sigma(X-\mu_x)\ (Z-\mu_z)$$

$$= \frac{2}{N\sigma_x}\ \Sigma(XZ - \mu_x Z - \mu_z X + \mu_x\mu_z)$$

$$= \frac{2}{N\sigma_x}\ [\ \Sigma XZ - \mu_x \Sigma Z - \mu_z \Sigma X + N\mu_x\mu_z\ ],$$

by the summation rules

$$= \frac{2}{N\sigma_x} [\Sigma XZ - \mu_x N\mu_z - \mu_z N\mu_x + N\mu_x\mu_z]$$

since $\Sigma Z = N\mu_z$ and $\Sigma X = N\mu_x$

$$= \frac{2}{N\sigma_x} [\Sigma XZ - N\mu_z\mu_x]$$

$$= \frac{2}{N\sigma_x} [\Sigma X_1 - N(\frac{1}{2}) \mu_x],$$

where ΣX_1 is the sum of X for group 1 and since the mean of a (1,0) uniform dichotomy is equal to 1/2,

$$= \frac{2}{N\sigma_x} [\frac{N}{2}\mu_{x_1} - \frac{N}{2}\mu_x],$$

since there are $N/2$ observations in group 1 and $\Sigma X_1 = (N/2)\,\mu_{x_1}$, where μ_{x_1} is the mean of X for group 1,

$$= \frac{1}{\sigma_x} (\mu_{x_1} - \mu_x) = \frac{x_{b_1}}{\sigma_x}$$

Similarly,

$$\rho_{yz} = \frac{1}{\sigma_y} (\mu_{y_1} - \mu_y) = \frac{y_{b_1}}{\sigma_y}.$$

Therefore,

$$\rho_{xy.z} = \frac{\eta_x\eta_y\rho_b + \sqrt{1-\eta_x^2}\sqrt{1-\eta_y^2}\ \rho_w - \dfrac{x_{b_1}\,y_{b_1}}{\sigma_x\sigma_y}}{\sqrt{1-\left(\dfrac{x_{b_1}}{\sigma_x}\right)^2}\sqrt{1-\left(\dfrac{y_{b_1}}{\sigma_y}\right)^2}}$$

But,

$$\eta_x^2 = \frac{\Sigma n_j\, x_{b_j}^2}{\Sigma x_t^2} = \frac{\frac{N}{2}\Sigma x_{b_j}^2}{N\sigma_x^2} = \frac{\frac{N}{2}\left(2x_{b_1}^{\ 2}\right)}{N\sigma_x^2} = \frac{x_{b1}^2}{\sigma_x^2}$$

by formula (6.5) and simplifying.

Similarly,

$$\eta_y^2 = \frac{{y_{b_1}}^2}{\sigma_y^2}.$$

Substituting and simplifying,

$$\rho_{xy.z} = \frac{\dfrac{x_{b_1} y_{b_1}}{\sigma_x \sigma_y} \rho_b + \sqrt{\dfrac{\sigma_x^2 - {x_{b_1}}^2}{\sigma_x^2}} \sqrt{\dfrac{\sigma_y^2 - {y_{b_1}}^2}{\sigma_y^2}} \rho_w - \dfrac{x_{b_1} y_{b_1}}{\sigma_x \sigma_y}}{\sqrt{\dfrac{\sigma_x^2 - {x_{b_1}}^2}{\sigma_x^2}} \sqrt{\dfrac{\sigma_y^2 - {y_{b_1}}^2}{\sigma_y^2}}}$$

But ρ_b must be either $+1$ or -1 (or indeterminate, in which case the whole thing blows up) since there are only two data points for the between-aggregate correlation. If it is $+1$, x_{b_1} and y_{b_1} are either both positive or both negative. If it is -1, x_{b_1} and y_{b_1} are of opposite sign. *For both cases* the first term and the third term in the numerator of the previous formula "kill each other," and the coefficient of σ_w is equal to the expression in the denominator, so that can be cancelled out.

Therefore,

$$\rho_{xy.z} = \rho_w.$$

UMAP

Modules in Undergraduate Mathematics and its Applications

Module 652

Spacecraft Attitude, Rotations and Quaternions

Dennis Pence

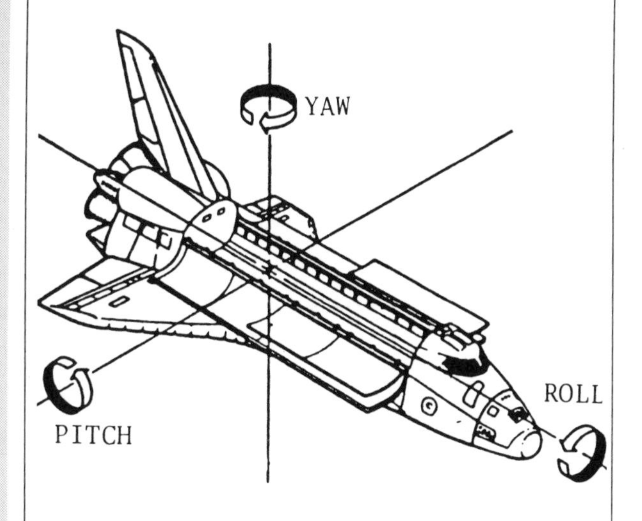

Published in cooperation with the Society for Industrial and Applied Mathematics, the Mathematical Association of America, the National Council of Teachers of Mathematics, the American Mathematical Association of Two-Year Colleges, and The Institute of Management Sciences.

INTERMODULAR DESCRIPTION SHEET:	UMAP Unit 652
TITLE:	SPACECRAFT ATTITUDE, ROTATIONS AND QUATERNIONS
AUTHOR:	Dennis Pence Department of Mathematics Western Michigan University Kalamazoo, MI 49008
MATH FIELD:	Linear Algebra
APPLICATION FIELD:	Physics (Space Flight)
TARGET AUDIENCE:	Students in a linear algebra course or a subsequent applications course.
ABSTRACT:	The rotational orientation of a spacecraft is called the attitude. Attitude can be represented by a matrix, a vector and an angle, or a sequence of angles. One of the more interesting representations uses quaternions. This module explores quaternion algebra and the relationships between these representations.
PREREQUISITES:	The following linear algebra skills: the ability to compute the norm, dot product, and cross product of three-dimensional vectors; to find the matrix representing a linear operator; understand orthonormalization and orthogonal matrices; and know how to compute determinants, eigenvalues, and eigenvectors.

The UMAP Journal, Vol. V, No. 2, 1984

Spacecraft Attitude, Rotations and Quaternions

Dennis Pence

Western Michigan University
Kalamazoo, MI 49008
(616) 383-6165

Table of Contents

1. INTRODUCTION 1
2. ATTITUDE MATRIX 1
 2.1 Definition 1
 2.2 Direction Cosine Matrix 2
 2.3 Transition Matrix 3
 2.4 Proper Orthogonal Matrix 3
 2.5 Rotation Operator 6
 2.6 Euler Angle Sequences 7
3. QUATERNION ALGEBRA 10
 3.1 Historical Development 10
 3.2 Vector Space Properties 11
 3.3 Quaternion Product 12
 3.4 Further Properties 14
4. QUATERNION ROTATION OPERATOR 16
 4.1 Definition 16
 4.2 Properties 16
 4.3 Representation for Attitude 19
5. KINEMATIC EQUATIONS FOR ATTITUDE 22
 5.1 Dynamics and Kinematics 22
 5.2 Quaternion Kinematics 22
 5.3 Attitude Matrix Kinematics 24
 5.4 Euler Angle Kinematics 25
6. BIBLIOGRAPHY 27
7. ACKNOWLEDGEMENT 27
8. MODEL EXAMINATION 28
9. ANSWERS TO EXERCISES 30
10. ANSWERS TO MODEL EXAM 38

Modules and Monographs in Undergraduate Mathematics and its Applications Project (UMAP)

The goal of UMAP was to develop, through a community of users and developers, a system of instructional modules in undergraduate mathematics and its applications to be used to supplement existing courses and from which complete courses may eventually be built.

The Project was guided by a National Advisory Board of mathematicians, scientists, and educators. UMAP was funded by a grant from the National Science Foundation and is now supported by the Consortium for Mathematics and Its Applications, Inc. (COMAP), a nonprofit corporation engaged in research and development in mathematics education.

This material was prepared with the partial support of National Science Foundation Grant No. SPE-8304192. Recommendations expressed are those of the authors and do not necessarily reflect the views of the NSF or the copyright holder.

1. Introduction

The motion of satellites is separated into two parts: orbit and attitude. The *orbit* (or more correctly the trajectory) is the path of the center of mass of the spacecraft. Orbit determination and prediction have a long history in mathematics. For example, Johann Kepler (1571-1630) discovered his famous laws describing orbits empirically. Knowing of these results, Isaac Newton (1642-1727) developed his laws of gravitation and motion, as well as much of the calculus, to further explain the orbital motion of planets and moons.

The *attitude* is the rotational orientation of the spacecraft about its center of mass. The earliest artificial satellites showed that, without some form of attitude control, a satellite begins to spin wildly. Now highly accurate knowledge and control are often critical for the mission of the spacecraft. Communication and earthsensing satellites need to have an antenna or camera pointing toward the earth. Solar-powered vehicles need arrays of photoelectric cells pointed toward the sun. Some of the ideas about rotations date back to Leonard Euler (1707-1783). Modern space travel has simply given the questions involving attitude new importance.

Spacecraft and even aircraft are able to take a variety of directional readings and to perform rotational maneuvers. The attitude of the craft is parametrized in such a way that this information can be interpreted in some fixed frame of reference. This module presents some of the frequently used representations for attitude, and derives the kinematic equations for rotational motion. We show why the representation using quaternions is often preferred for space attitude analysis.

2. Attitude Matrix

2.1 Definition

Suppose that we have a fixed *reference frame* giving a coordinate system with the standard basis $\vec{i}$, $\vec{j}$, and $\vec{k}$ in $I\!R^3$. Consider a rigid body, for example a satellite, with a locally defined coordinate system called the *body system* given by the right-handed triad of unit vectors $\vec{u}$, $\vec{v}$, and $\vec{w}$. In the reference frame,

$$\vec{u} = u_1\vec{i} + u_2\vec{j} + u_3\vec{k},\ v = v_1\vec{i} + v_2\vec{j} + v_3\vec{k},$$

and

$$\vec{w} = w_1\vec{i} + w_2\vec{j} + w_3\vec{k}$$

with

$$\vec{u} \cdot \vec{u} = \vec{v} \cdot \vec{v} = \vec{w} \cdot \vec{w} = 1,$$

$$\vec{u} \cdot \vec{v} = \vec{v} \cdot \vec{w} = \vec{w} \cdot \vec{u} = 0,$$

$$\vec{u} \times \vec{v} = \vec{w}.$$

(Note that a left-handed triad would have instead $\vec{u} \times \vec{v} = -\vec{w}$.) See, for example, [Anton, Chapter 3].

We define the *attitude matrix* of the body to be

$$A = \begin{bmatrix} u_1 & u_2 & u_3 \\ v_1 & v_2 & v_3 \\ w_1 & w_2 & w_3 \end{bmatrix}$$

2.2 Direction Cosine Matrix

The attitude matrix A can be considered as a direction cosine matrix. Each element is the cosine of the angle between one of the body unit vectors and one of the standard basis vectors. For example

$$u_3 = \cos \alpha_3 = \frac{\vec{u} \cdot \vec{k}}{|\vec{u}|\ |\vec{k}|}.$$

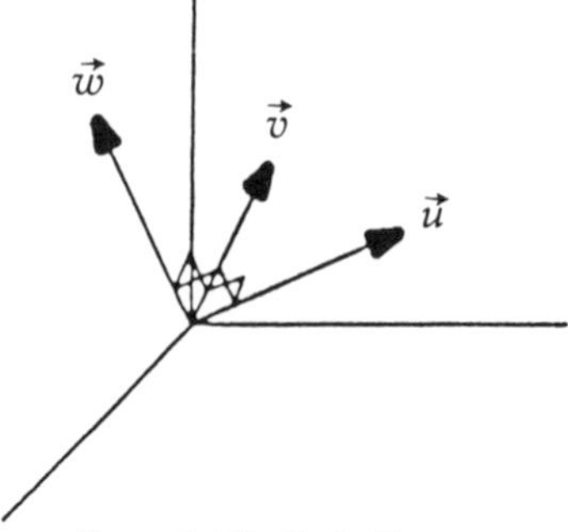

Figure 1. The Body System.

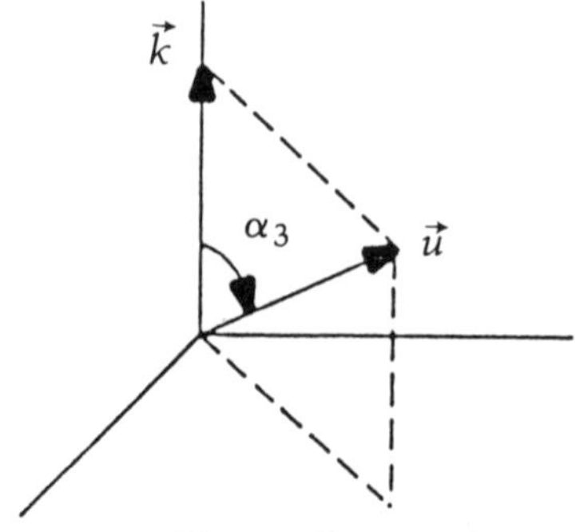

Figure 2. Direction Cosines.

2.3 Transition Matrix

We can easily see that the attitude matrix A is the transition matrix from the standard basis $\{\vec{i}, \vec{j}, \vec{k}\}$ to the body system $\{\vec{u}, \vec{v}, \vec{w}\}$, i.e, if $\vec{a} = a_1\vec{i} + a_2\vec{j} + a_3\vec{k}$ and

$$\begin{bmatrix} b_1 \\ b_2 \\ b_3 \end{bmatrix} = \begin{bmatrix} u_1 & u_2 & u_3 \\ v_1 & v_2 & v_3 \\ w_1 & w_2 & w_3 \end{bmatrix} \begin{bmatrix} a_1 \\ a_2 \\ a_3 \end{bmatrix}$$

then $\vec{a} = b_1\vec{u} + b_2\vec{v} + b_3\vec{w}$. See [Anton, Section 4.10]. This transition from one basis to the other can be important because a direction reading from the satellite is measured in the body system while the result may need to be interpreted in the reference frame or vice versa.

2.4 Proper Orthogonal Matrix

Recall that a square matrix P is called *orthogonal* if the rows (or the columns) of P form an orthonormal set, or equivalently, if

$$P\,P^t = I.$$

Thus it is easy to find the inverse of an orthogonal matrix,

$$P^{-1} = P^t.$$

See [Anton, Section 4.10].

Since the rows of the attitude matrix A come from the right-hand triad of orthonormal vectors, A is orthogonal. A^t then is the transition matrix from the basis $\{\vec{u}, \vec{v}, \vec{w}\}$ to the standard basis.

For any orthogonal matrix P,

$$1 = \det I = \det P \det P^t = (\det P)^2.$$

Thus the determinant of an orthogonal matrix is either $+1$ or -1.

Using the scalar triple product,

$$\det A = \begin{vmatrix} u_1 & u_2 & u_3 \\ v_1 & v_2 & v_3 \\ w_1 & w_2 & w_3 \end{vmatrix}$$

$$= \vec{u} \cdot (\vec{v} \times \vec{w}) = \vec{u} \cdot \vec{u} = 1.$$

Such an orthogonal matrix with determinant 1 is called *proper.* (Note that a left-handed triad would give an improper orthogonal direction cosine matrix with determinant -1.)

THEOREM 2.1. The number $\lambda = 1$ is an eigenvalue for any 3×3 proper orthogonal matrix A, i.e., there exists a nonzero vector $\vec{n} = n_1\vec{i} + n_2\vec{j} + n_3\vec{k}$ with

$$A \begin{bmatrix} n_1 \\ n_2 \\ n_3 \end{bmatrix} = \begin{bmatrix} n_1 \\ n_2 \\ n_3 \end{bmatrix}$$

PROOF. Recall for $z = x + yi \in \mathbb{C}$, the conjugate is $z^* = x - yi$. The conjugate of a column vector or of a matrix with complex entries is found by taking the conjugate of every entry. The norm (or modulus) is

$$|z| = \sqrt{z^*z} \quad \text{for } z \in \mathbb{C}$$

$$\left\| \begin{bmatrix} a \\ b \\ c \end{bmatrix} \right\| = \sqrt{\begin{bmatrix} a \\ b \\ c \end{bmatrix}^{*t} \begin{bmatrix} a \\ b \\ c \end{bmatrix}} = \sqrt{|a|^2 + |b|^2 + |c|^2}$$

$$\text{for } \begin{bmatrix} a \\ b \\ c \end{bmatrix} \in \mathbb{C}^3.$$

Multiplication by a real orthogonal matrix A preserves the norm because

$$\left\| A \begin{bmatrix} a \\ b \\ c \end{bmatrix} \right\|^2 = \left\{ A \begin{bmatrix} a \\ b \\ c \end{bmatrix} \right\}^{*t} A \begin{bmatrix} a \\ b \\ c \end{bmatrix}$$

$$= (a^* \ b^* \ c^*) \ A^t \ A \begin{bmatrix} a \\ b \\ c \end{bmatrix} = \left\| \begin{bmatrix} a \\ b \\ c \end{bmatrix} \right\|^2$$

Suppose $\lambda \in \mathbb{C}$ is an eigenvalue for A with eigenvector $(a \ b \ c)^t$. See [Anton, Chapter 6]. Then

$$\left\| \lambda \begin{bmatrix} a \\ b \\ c \end{bmatrix} \right\| = \left\| A \begin{bmatrix} a \\ b \\ c \end{bmatrix} \right\| = \left\| \begin{bmatrix} a \\ b \\ c \end{bmatrix} \right\|$$

which implies that $|\lambda| = 1$. Eigenvalues are roots of the characteristic polynomial

$$p(\lambda) = \det(\lambda I - A) = (\lambda - \lambda_1)(\lambda - \lambda_2)(\lambda - \lambda_3),$$

where p has real coefficients since A is real. Any complex eigenvalues must come in conjugate pairs. Assuming that A is proper,

$$p(0) = \det(-A) = -\lambda_1 \lambda_2 \lambda_3 = -1$$

or

$$\lambda_1 \lambda_2 \lambda_3 = 1.$$

Ignoring ordering, there are only three possibilities for the eigenvalues of A.

(i) $\lambda_1 = \lambda_2 = \lambda_3 = 1$.

(ii) If one root is complex, say λ_1, then another must be the conjugate, say $\lambda_2 = \lambda_1^*$, and

$$1 = \lambda_1 \lambda_2 \lambda_3 = \lambda_1 \lambda_1^* \lambda_3 = \lambda_3.$$

(iii) If no root is complex and one equals -1, say λ_1, then

$$1 = \lambda_1 \lambda_2 \lambda_3 = (-1) \lambda_2 \lambda_3 = \lambda_3.$$

In every case $\lambda = 1$ is an eigenvalue for A.

2.5 Rotation Operator

We can interpret the linear operator R_A: $IR^3 \rightarrow IR^3$ defined by multiplication with a proper orthogonal matrix A as a *rotation.* See [Anton, Section 5.3]. Start with a unit eigenvector

$$\vec{n} = n_1\vec{i} + n_2\vec{j} + n_3\vec{k}$$

associated with the eigenvalue $\lambda = 1$. Add two independent vectors and apply the Gram-Schmidt process to obtain an orthonormal basis $\{\vec{n}, \vec{l}, \vec{m}\}$. Taking the negative of $\vec{m}$ if necessary, we have a right-hand triad of unit vectors $\{\vec{l}, \vec{m}, \vec{n}\}$.

Let B be the matrix representing R_A with respect to the basis $\{\vec{l}, \vec{m}, \vec{n}\}$. If P is the proper orthogonal transition matrix from the standard basis to $\{\vec{l}, \vec{m}, \vec{n}\}$, then

$$B = P\ A\ P^t$$

$$BB^t = P\ A\ P^t\ P\ A^t\ P^t = I,$$

$$\det\ B = \det P\ \det A\ \det P^t = 1.$$

The fact that $R_A(\vec{n}) = \vec{n}$ implies that multiplication of $(0\ \ 0\ \ 1)^t$ by B leaves it unchanged. Combining all of these properties, we must have B in the following form:

$$B = \begin{bmatrix} \cos\Phi & \sin\Phi & 0 \\ -\sin\Phi & \cos\Phi & 0 \\ 0 & 0 & 1 \end{bmatrix}$$

This matrix represents a rotation about the axis indicated by the vector n through an angle Φ, where the sense of the angle of rotation is indicated by the right-hand rule: pointing the thumb of the right hand in the direction n, the fingers curl in the direction of a positive angle Φ. This is summarized in the following which is called *Euler's Theorem*.

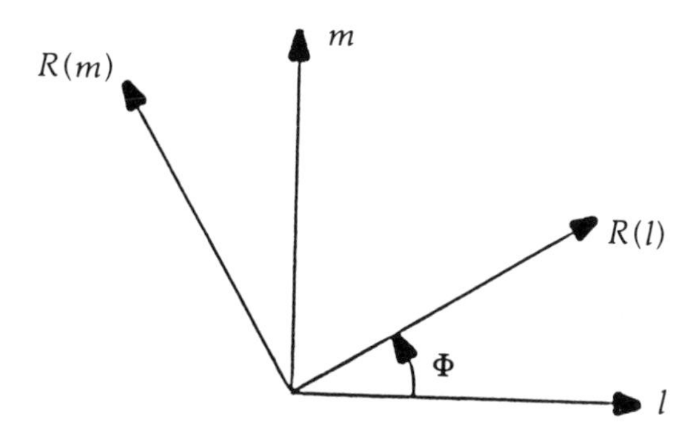

Figure 3. Convention for Rotation Angle Measurement (n is pointing out of the paper).

THEOREM 2.2. The most general displacement of a rigid body with one fixed point (for us the origin) is a rotation about some axis.

2.6 Euler Angle Sequence

Usually it is hard to identify what the axis of rotation, $\vec{n}$, will be for the attitude matrix A. However, if the axis of rotation is one of the standard basis vectors, the attitude matrix takes a simple form. Namely, with axis $\vec{i}$, $\vec{j}$, and $\vec{k}$ (respectively) we get

$$A_1(\Phi) = \begin{bmatrix} 1 & 0 & 0 \\ 0 & \cos\Phi & \sin\Phi \\ 0 & -\sin\Phi & \cos\Phi \end{bmatrix}$$

$$A_2(\Phi) = \begin{bmatrix} \cos\Phi & 0 & -\sin\Phi \\ 0 & 1 & 0 \\ \sin\Phi & 0 & \cos\Phi \end{bmatrix}$$

$$A_3(\Phi) = \begin{bmatrix} \cos\Phi & \sin\Phi & 0 \\ -\sin\Phi & \cos\Phi & 0 \\ 0 & 0 & 1 \end{bmatrix}$$

Airplane pilots frequently want to rotate their craft in the above three ways. Sitting in the pilot's seat, the x-axis is forward and backward, the y-axis is right and left, and the z-axis is up and down. A *roll* is a rotation through an angle ϕ about the x-axis ($\vec{n} = \vec{i}$), a *pitch* is a rotation through an angle θ about the y-axis ($\vec{n} = \vec{j}$), and a *yaw* is a rotation through an angle ψ about the z-axis ($\vec{n} = \vec{k}$). Since all of the astronauts have been pilots, the same terminology is used in space flight. One frequently hears that a command module or shuttle is making a "yaw maneuver" which means it is being rotated about the z-axis.

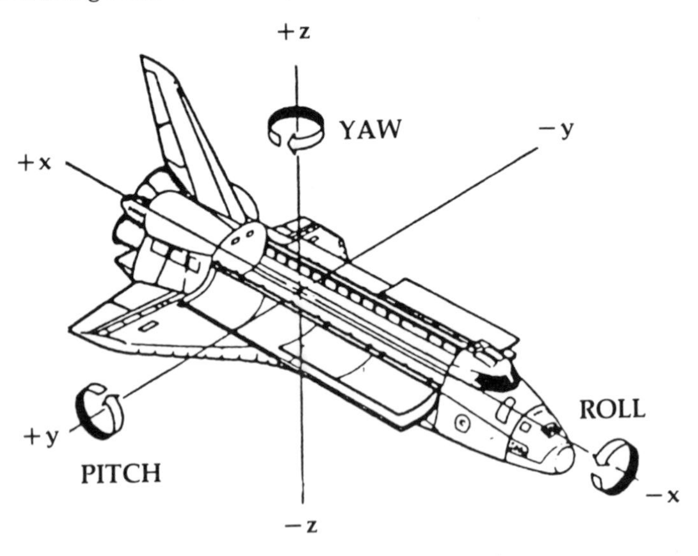

Figure 4. The orbiter's three axes. When the orbiter is gravity-stabilized, its nose always points earthward. When the orbiter drifts quasi-inertially, one surface constantly faces away from the sun to shade the heat radiators. (Source: *Air and Space*, Volume 2, No. 3, January-February, 1979, page 6.)

An arbitrary rotation can be obtained as a composition of these special rotations about the coordinate axes. For example

$$A_{3\text{-}1\text{-}2} = A_2(\theta)\ A_1(\phi)\ A_3(\psi)$$

since composition is obtained by successive matrix multiplication. This is called a 3-1-2 *Euler angle sequence*. The first applied operator can be any of the three, the middle must be different from the first, and the last must be different from the middle so that there are 12 possible Euler angle sequences. For instance, the 3-1-3 sequence is described in [Goldstein]. Different orderings give genuinely different results because matrix multiplication is not commutative.

For simplicity, we will consider only the 3-1-2 sequence. Notice that the angles ψ, ϕ, and θ give a parametrization for an attitude matrix $A_{3\text{-}1\text{-}2}$. The reader is encouraged to carry out the multiplications indicated above to see how complicated this representation is.

Exercises

2.1. Given

$$\vec{u} = \frac{1}{\sqrt{2}}\vec{i} + \frac{1}{\sqrt{2}}\vec{j}$$

and

$$\vec{v} = \frac{-1}{\sqrt{10}}\vec{i} + \frac{1}{\sqrt{10}}\vec{j} + \frac{2}{\sqrt{5}}\vec{k},$$

find $\vec{w}$ and the attitude matrix A. Verify directly that $AA^t = I$.

2.2. Find a 3×3 orthogonal matrix which is *not* proper. Then show that the linear operator which it represents transforms any right-hand triad into a left-hand triad (as in a mirror reflection).

2.3. The steps given in Section 2.5 are reversible. Starting with $\vec{n}$ and Φ, the axis and angle of rotation, show that the following formula describes the attitude matrix.

$$A = \cos \Phi \, I + (1 - \cos \Phi) \begin{bmatrix} n_1 \\ n_2 \\ n_3 \end{bmatrix} (n_1 \, n_2 \, n_3) + \sin \Phi \begin{bmatrix} 0 & +n_3 & -n_2 \\ -n_3 & 0 & +n_1 \\ +n_2 & -n_1 & 0 \end{bmatrix}.$$

Case 1: Assume $n_1{}^2 + n_2{}^2 \neq 0$. Select

$$\vec{l} = (n_2\vec{i} - n_1\vec{j}) \, |n_2\vec{i} - n_1\vec{j}|^{-1}$$

and

$$\vec{m} = \vec{n} \times \vec{l} \; A = P^t B P$$

in the notation of the section.

Case 2: When $n_1{}^2 + n_2{}^2 = 0$, finding A is even easier.

2.4. Let $A = A_{3\text{-}1\text{-}2} = A_1(\phi) \, A_3(\psi)$ where $\psi = \pi/4$, $\phi = \pi/3$, and $\theta = 0$. Find $\vec{n}$ and Φ, the axis and angle of rotation.

2.5. A rotation is not uniquely represented by $\vec{n}$ and Φ, an axis and angle of rotation. Show that $-\vec{n}$ and $(2\pi - \Phi)$ represents the same rotation.

3. Quaternion Algebra

3.1 Historical Development

Sir William Rowan Hamilton (1805-1865) was a very famous mathematician—known primarily for his work in the calculus of variations and applications to dynamics and optics—before he discovered quaternions. He had been trying for some time to generalize complex numbers. On October 16, 1843 as he was walking to a meeting of the Council of the Royal Irish Academy, he suddenly realized what was needed. He was so excited that he carved the fundamental identity,

$$i^2 = j^2 = k^2 = ijk = -1,$$

in the stone of the Broughman Bridge. He also announced the result later at the meeting.

The value of this discovery has been disputed ever since that day. Hamilton spent the remaining 22 years of his life working with quaternions. His results can be found in his book, *Elements of Quaternions,* which was published in 1866, one year after his death. Those who see little merit in this area of mathematics might say that he wasted those last years. See [E.T. Bell, *Men of Mathematics*], for example. Perhaps a more balanced judgment is to consider how these ideas involving quaternions, due to Hamilton and others, led to modern vector analysis. (You are strongly urged to browse in the delightful book [M.J. Crow, *A History of Vector Analysis*].) The American Josiah Willard Gibbs (1839-1903) and the Englishman Oliver Heaviside (1850-1925) are credited with initiating modern vector algebra in the 1880s. Vectors are now used extensively in mathematics, physics, and engineering. Interest in quaternion algebra rapidly diminished when vector algebra appeared.

This module presents one application of quaternions which has survived, namely in satellite attitude work. Quaternions are almost always used in some phases of the design and analysis of attitude control systems. Often, quaternions are the parameters used in the ground or onboard computer calculations involving attitude. See [J.R. Wertz, *Spacecraft Attitude Determination and Control*].

There are also important modern applications of quaternion algebra to geometry and group theory. See [H.S.M. Coxeter, *Regular Complex Polytopes*].

3.2 Vector Space Properties

Let the elements of the space H of quaternions be identical to the elements of Euclidean four-dimensional space IR^4, with the usual definitions of addition, subtraction, multiplication by a scalar and norm. We will denote a typical element either as a column vector,

$$\mathbf{q} = \begin{bmatrix} q_1 \\ q_2 \\ q_3 \\ q_4 \end{bmatrix} \text{ and } \mathbf{i} = \begin{bmatrix} 1 \\ 0 \\ 0 \\ 0 \end{bmatrix}, \mathbf{j} = \begin{bmatrix} 0 \\ 1 \\ 0 \\ 0 \end{bmatrix}, \mathbf{k} = \begin{bmatrix} 0 \\ 0 \\ 1 \\ 0 \end{bmatrix}, \mathbf{l} = \begin{bmatrix} 0 \\ 0 \\ 0 \\ 1 \end{bmatrix}$$

or as a linear combination of the special basis given above,

$$\mathbf{q} = q_1\mathbf{i} + q_2\mathbf{j} + q_3\mathbf{k} + q_4\mathbf{l}.$$

We call q_4 (or more precisely $q_4\mathbf{l}$) the *scalar part* of the quaternion $\mathbf{q}$ and also denote this by $S(\mathbf{q})$. Usually the basis element $\mathbf{l}$ is omitted. The rest, $\vec{q} = q_1\mathbf{i} + q_2\mathbf{j} + q_3\mathbf{k}$, is called the *vector part* of $\mathbf{q}$. This is also denoted by $V(q)$. Most of this notation and the term *vector* are due to Hamilton. We can think of IR^3 as embedded in H in a natural way as the set of all quaternions with zero scalar part. Much of modern vector algebra can be found by considering vectors and scalars separately, rather than as a part of one quantity.

3.3 Quaternion Product

We now define the product of two elements in H. This is most easily done by first giving all possible products involving two basis elements.

$$\mathbf{ii} = \mathbf{i}^2 = -1, \qquad \mathbf{ij} = \mathbf{k} = -\mathbf{ji},$$

$$\mathbf{jj} = \mathbf{j}^2 = -1, \qquad \mathbf{jk} = \mathbf{i} = -\mathbf{kj},$$

$$\mathbf{kk} = \mathbf{k}^2 = -1, \qquad \mathbf{ki} = \mathbf{j} = -\mathbf{ik},$$

and

$$\mathbf{lq} = \mathbf{ql} = \mathbf{q} \qquad \text{for any } \mathbf{q} \in H.$$

We can see in the above list, the fundamental identity which Hamilton carved in stone. Restricting attention to just $\{\pm\mathbf{i}, \pm\mathbf{j}, \pm\mathbf{k}, \pm\mathbf{l}\}$, we have the finite quaternion group with the quaternion product.

This product is extended to all of H by linearity, i.e., we require that

$$\mathbf{p}(\mathbf{q} + \mathbf{r}) = \mathbf{pq} + \mathbf{pr},$$

$$(\mathbf{p} + \mathbf{q})\mathbf{r} = \mathbf{pr} + \mathbf{qr},$$

$$(\alpha\mathbf{p})\mathbf{q} = \mathbf{p}(\alpha\mathbf{q}) = (\mathbf{pq}), \text{ for } \alpha \in IR.$$

Example.

$$(2\mathbf{i} - 3\mathbf{j} + 5)(\mathbf{j} + 4\mathbf{k})$$

$$= 2\mathbf{ij} - 3\mathbf{jj} + 5\mathbf{j} + 8\mathbf{ik} - 12\mathbf{jk} + 20\mathbf{k}$$

$$= -12\mathbf{i} - 3\mathbf{j} + 22\mathbf{k} + 3.$$

PROPOSITION 3.1. Suppose $\mathbf{p} = \vec{p} + p_4$, $\mathbf{q} = \vec{q} + q_4$.

(a) $S(\mathbf{pq}) = p_4 q_4 - \vec{p} \cdot \vec{q}$.

(b) $V(\mathbf{pq}) = p_4 \vec{q} + q_4 \vec{p} + \vec{p} \times \vec{q}$.

Using this result, we note that in the special case when $\mathbf{p}$ and $\mathbf{q}$ are vectors (with zero scalar parts), we have

$$S(\vec{p}\vec{q}) = -\vec{p} \cdot \vec{q}, \qquad V(\vec{p}\vec{q}) = \vec{p} \times \vec{q}.$$

The familiar vector dot product and vector cross product were thus implied and anticipated in the quaternion product.

PROPOSITION 3.2. The quaternion product has the following matrix multiplication interpretation.

$$\mathbf{pq} = \begin{bmatrix} q_4 & q_3 & -q_2 & q_1 \\ -q_3 & q_4 & q_1 & q_2 \\ q_2 & -q_1 & q_4 & q_3 \\ -q_1 & -q_2 & -q_3 & q_4 \end{bmatrix} \begin{bmatrix} p_1 \\ p_2 \\ p_3 \\ p_4 \end{bmatrix}$$

Since we will have several further opportunities to use the skew-symmetric matrix in Proposition 3.2, we define

$$\widetilde{Q}(\mathbf{q}) = \begin{bmatrix} q_4 & q_3 & -q_2 & q_1 \\ -q_3 & q_4 & q_1 & q_2 \\ q_2 & -q_1 & q_4 & q_3 \\ -q_1 & -q_2 & -q_3 & q_4 \end{bmatrix}$$

Note that in Exercise 2.3 we saw the principal minor of $\tilde{Q}(n)$ obtained by deleting the last row and column. More generally, we denote this minor by

$$Q(\mathbf{q}) = \begin{bmatrix} q_4 & q_3 & -q_2 \\ -q_3 & q_4 & q_1 \\ q_2 & -q_1 & q_4 \end{bmatrix}$$

3.4 Further Properties

In analogy with complex numbers, we say

$$\mathbf{q}^* = -q_1\mathbf{i} - q_2\mathbf{j} - q_3\mathbf{k} + q_4$$

is the *conjugate* of the quaternion

$$\mathbf{q} = q_1\mathbf{i} + q_2\mathbf{j} + q_3\mathbf{k} + q_4.$$

PROPOSITION 3.3. For any $\mathbf{p}$ and $\mathbf{q}$ in H we have

(a) $\mathbf{q}\mathbf{q}^* = \mathbf{q}^*\mathbf{q} = |\mathbf{q}|^2$,

(b) $\mathbf{q}^{**} = \mathbf{q}$,

(c) $(\mathbf{p}\mathbf{q})^* = \mathbf{q}^*\mathbf{p}^*$,

(d) $|\mathbf{p}\mathbf{q}| = |\mathbf{p}|\ |\mathbf{q}|$.

Using property (**a**) above, we can find the *inverse* for any quaternion $\mathbf{q}$, namely

$$\mathbf{q}^{-1} = \frac{1}{|\mathbf{q}|^2}\mathbf{q}^*.$$

Note that the inverse of a unit quaternion is its conjugate. Division is not defined, however, because $\mathbf{p}\mathbf{q}^{-1}$ and $\mathbf{q}^{-1}\mathbf{p}$ are not, in general, the same. Remember that the quaternion product is not commutative.

Examples.

$$(5\mathbf{i} + 3\mathbf{k})\,(2\mathbf{j} + \mathbf{k} + 1) = -\mathbf{i} - 5\mathbf{j} + 13\mathbf{k} - 3$$

$$(2\mathbf{j} + \mathbf{k} + 1)\,(5\mathbf{i} + 3\mathbf{k}) = 11\mathbf{i} + 5\mathbf{j} - 7\mathbf{k} - 3$$

$$(2\mathbf{j} + \mathbf{k} + 1)\,((-2\mathbf{j} - 2\mathbf{k} + 1)/6)$$

$$= ((-2\mathbf{j} - 2\mathbf{k} + 1)/6)\,(2\mathbf{j} + \mathbf{k} + 1) = 1.$$

Exercises

3.1. Perform the indicated operations.

(a) $(\mathbf{i} - 5\mathbf{j} + \mathbf{k} + 2) + (3\mathbf{j} + 7\mathbf{k} - 4)$.

(b) $(2\mathbf{i} + 3\mathbf{j} - \mathbf{k} + 1)\,(\mathbf{i} + 4\mathbf{j} + \mathbf{k} - 2)$.

(c) $(\mathbf{i} - 5\mathbf{j} + 2\mathbf{k} + 3)^2$.

(d) $|5\mathbf{i} - \mathbf{j} + 10|$.

3.2. Prove Proposition 3.1.

3.3. Verify Proposition 3.2 by showing the results hold when multiplying two basis elements and then appealing to linearity.

3.4. Observe that the *complex numbers* can be naturally considered as the subspace of H consisting of quaternions q where $q_2 = q_3 = 0$. Verify that the norm and the product in H give the usual modulus and complex product when restricted to this subspace.

3.5. Prove Proposition 3.3, part (c).

3.6. Show that $\mathbf{q}^2 = -1$ if and only if $|\mathbf{q}| = 1$ and $S(\mathbf{q}) = 0$.

4. Quaternion Rotation Operator

4.1 Definition

Given a fixed quaternion $\mathbf{q}$, we define an operator $M_{\mathbf{q}}: H \to H$ by

$$M_{\mathbf{q}}(\mathbf{p}) = \mathbf{q}^{-1}\mathbf{p}\mathbf{q}.$$

Using the formula for the inverse and the linearity of the quaternion product, we have

$$\mathbf{q}^{-1}\mathbf{p}\mathbf{q} = \frac{1}{|\mathbf{q}|^2}\mathbf{q}^*\mathbf{p}\mathbf{q} = \frac{\mathbf{q}^*}{|\mathbf{q}^*|}\mathbf{p}\frac{\mathbf{q}}{|\mathbf{q}|}.$$

For this reason, it suffices to only consider unit quaternions (with $|\mathbf{q}| = 1$) where the formula simplifies to

$$M_{\mathbf{q}}(\mathbf{p}) = \mathbf{q}^*\mathbf{p}\mathbf{q}.$$

Notice that we can apply $M_{\mathbf{q}}$ to a vector $\vec{p} \in I\!R^3$ by considering the vector as a quaternion $\mathbf{p} = \vec{p}$ with zero scalar part. Arthur Cayley (1821-1895) is credited with introducing this operator and giving its connection with rotations.

4.2 Properties

The linearity of the quaternion product implies that

$$M_{\mathbf{q}}(a\mathbf{p} + b\mathbf{r}) = aM_{\mathbf{q}}(\mathbf{p}) + bM_{\mathbf{q}}(\mathbf{r}).$$

THEOREM 4.1. LET $\mathbf{p}$ and $\mathbf{q}$ be unit quaternions.

(a) $M_{\mathbf{p}} \circ M_{\mathbf{q}} = M_{\mathbf{qp}}$. (Composition)

(b) $(M_{\mathbf{q}})^{-1} = M_{\mathbf{q}^*}$. (Inverse Operator)

PROOF.

(a) $M_{\mathbf{p}}(M_{\mathbf{q}}(\mathbf{r})) = \mathbf{p}^*(\mathbf{q}^*\mathbf{r}\mathbf{q})\mathbf{p}$

$= (\mathbf{qp})^*\mathbf{r}(\mathbf{qp}) = M_{\mathbf{qp}}(\mathbf{r})$, for all $\mathbf{r} \in H$.

(b) Applying part (a) and that $\mathbf{qq}^* = \mathbf{q}^*\mathbf{q} = 1$,

$(M_{\mathbf{q}^*} \circ M_{\mathbf{q}})(\mathbf{r}) = (M_{\mathbf{q}} \circ M_{\mathbf{q}^*})(\mathbf{r}) = \mathbf{1r1} = \mathbf{r}.$

THEOREM 4.2. The subspace of pure vectors is invariant under the operator $M_{\mathbf{q}}$. Thus

$M_{\mathbf{q}}(\vec{p}) \in IR^3$ for any $\vec{p} \in IR^3$.

PROOF. We find the scalar part of $M_{\mathbf{q}}(\vec{p})$ by using Proposition 3.1 twice.

$$(\vec{p}\mathbf{q}) = (-\vec{p} \cdot \vec{q}) + (q_4\vec{p} + \vec{p} \times \vec{q})$$

$$S(\mathbf{q}^*(\vec{p}\mathbf{q})) = q_4(-\vec{p} \cdot \vec{q}) - (-\vec{q}) \cdot (q_4\vec{p} + \vec{p} \times \vec{q})$$

$$= -q_4(\vec{p} \cdot \vec{q}) + q_4(\vec{q} \cdot \vec{p}) + \vec{q} \cdot (\vec{p} \times \vec{q}).$$

The first two terms add to give zero and the last term is the scalar triple product which can be rewritten as a determinant.

$$S(\mathbf{q}^*\vec{\mathbf{p}}\mathbf{q})) = \begin{vmatrix} q_1 & q_2 & q_3 \\ p_1 & p_2 & p_3 \\ q_1 & q_2 & q_3 \end{vmatrix} = 0.$$

Thus $M_{\mathbf{q}}(\vec{p})$ is also a pure vector.

Since $M_{\mathbf{q}}$ is a linear operator on H, we can easily find the matrix which represents $M_{\mathbf{q}}$ with respect to the basis $\{\mathbf{i}, \mathbf{j}, \mathbf{k}, \mathbf{1},\}$.

$$\widetilde{A} = [M_{\mathbf{q}}(\mathbf{i}), M_{\mathbf{q}}(\mathbf{j}), M_{\mathbf{q}}(\mathbf{k}), M_{\mathbf{q}}(\mathbf{1})].$$

Using Proposition 3.2 and that $|q| = 1$, we have

$$\widetilde{A} = [\widetilde{Q}(\mathbf{q})[\mathbf{q}^*\mathbf{i}], \widetilde{Q}(\mathbf{q})[\mathbf{q}^*\mathbf{j}], \widetilde{Q}(\mathbf{q})[\mathbf{q}^*\mathbf{k}], \widetilde{Q}(\mathbf{q})[\mathbf{q}^*\mathbf{1}]]$$

$$= \widetilde{Q}(\mathbf{q})[\mathbf{q}^*\mathbf{i}, \mathbf{q}^*\mathbf{j}, \mathbf{q}^*\mathbf{k}, \mathbf{q}^*\mathbf{1}]$$

$$= \begin{bmatrix} q_4 & q_3 & -q_2 & q_1 \\ -q_3 & q_4 & q_1 & q_2 \\ q_2 & -q_1 & q_4 & q_3 \\ -q_1 & -q_2 & -q_3 & q_4 \end{bmatrix} \begin{bmatrix} q_4 & q_3 & -q_2 & -q_1 \\ -q_3 & q_4 & q_1 & -q_2 \\ q_2 & -q_1 & q_4 & -q_3 \\ q_1 & q_2 & q_3 & q_4 \end{bmatrix}$$

$$= \begin{bmatrix} (q_1^2-q_2^2-q_3^2+q_4^2) & 2(q_1q_2+q_3q_4) & 2(q_1q_3-q_2q_4) & 0 \\ 2(q_1q_2-q_3q_4) & (-q_1^2+q_2^2-q_3^2+q_4^2) & 2(q_2q_3+q_1q_4) & 0 \\ 2(q_1q_3+q_2q_4) & 2(q_2q_3-q_1q_4) & (-q_1^2-q_2^2+q_3^2+q_4^2) & 0 \\ 0 & 0 & 0 & 1 \end{bmatrix}.$$

Let A be the principal minor of $\widetilde{A}$ obtained by deleting the last row and column. Thus

$$M_{\mathbf{q}}(\mathbf{p}) = \widetilde{A}[\mathbf{p}] = \begin{bmatrix} A & 0 \\ 0 & 1 \end{bmatrix} \begin{bmatrix} \vec{p} \\ p_4 \end{bmatrix}.$$

We note that

$$M_{\mathbf{q}^*}(\mathbf{p}) = \widetilde{A}^t[\mathbf{p}] = \begin{bmatrix} A^t & 0 \\ 0 & 1 \end{bmatrix} \begin{bmatrix} \vec{p} \\ p_4 \end{bmatrix}.$$

Since $M_{\mathbf{q}}$ and $M_{\mathbf{q}^*}$ are inverse operators

$$\widetilde{A}\widetilde{A}^t = \widetilde{A}^t\widetilde{A} = I_4.$$

Given the block form of $\widetilde{A}$, this implies that

$$AA^t = A^tA = I_3.$$

Thus both $\widetilde{A}$ and A are orthogonal matrices.

PROPOSITION 4.1. In the notation above,

$$A = (q_4^2 - |\vec{q}|)I_3 + 2\begin{bmatrix} q_1 \\ q_2 \\ q_3 \end{bmatrix}(q_1\ q_2\ q_3) - 2q_4\,Q(\vec{q}).$$

4.3 Representation for Attitude

We now look at how the operator $M_{\mathbf{q}}$, considered only as an operator on pure vectors, describes attitude. For $\mathbf{p} = \vec{p}$,

$$M_{\mathbf{q}}(\vec{p}) = \begin{bmatrix} A & 0 \\ 0 & 1 \end{bmatrix} \begin{bmatrix} \vec{p} \\ 0 \end{bmatrix} = \begin{bmatrix} A\vec{p} \\ 0 \end{bmatrix} = \begin{bmatrix} R_A(\vec{p}) \\ 0 \end{bmatrix}$$

where A is orthogonal. We need only check the determinant of A.

$$\widetilde{A} = \left[\begin{array}{c|c} M_{\mathbf{q}}(\vec{i}),\ M_{\mathbf{q}}(\vec{j}),\ M_{\mathbf{q}}(\vec{k}) & 0 \\ \hline 0 & 1 \end{array}\right].$$

Using the scalar triple product and then Proposition 3.1

$$\begin{aligned} \det A &= M_{\mathbf{q}}(\vec{i}) \cdot (M_{\mathbf{q}}(\vec{j}) \times M_{\mathbf{q}}(\vec{k})) \\ &= M_{\mathbf{q}}(\vec{i}) \cdot V(M_{\mathbf{q}}(\vec{j})\ M_{\mathbf{q}}(\vec{k})) \\ &= M_{\mathbf{q}}(\vec{i}) \cdot V(\mathbf{q}^*\vec{j}\,\mathbf{q}\mathbf{q}^*\ \vec{k}\mathbf{q}) \\ &= M_{\mathbf{q}}(\vec{i}) \cdot V(M_{\mathbf{q}}(\vec{i})) = M_{\mathbf{q}}(\vec{i}) \cdot M_{\mathbf{q}}(\vec{i}) = 1. \end{aligned}$$

We have moved freely between quaternion and vector algebra, and we have used the fact that multiplication by an orthogonal matrix is norm-preserving so that $|M_{\mathbf{q}}(\vec{i})| = 1$.

We consider any unit quaternion $\mathbf{q}$ as a parameterization for an attitude matrix A via the operator $M_{\mathbf{q}}$. Notice that this attitude matrix can be obtained algebraically from the parameters in $\mathbf{q}$ with no trigonometric function evaluations. The previous representations using the axis/angle or using one of the Euler angle sequences involve trigonometric evaluations.

THEOREM 4.3. The unit quaternion $\mathbf{q} = \cos\beta + \vec{n}\sin\beta$ where $|\vec{n}| = 1$ represents the rotation operator with axis of rotation n and angle of rotation $\Phi = 2\beta$.

PROOF. It suffices to check the action of M_q on a right-hand triad $\{\vec{l}, \vec{m}, \vec{n}\}$. We claim

$$M_q(\vec{l}) = \cos(2\beta)\vec{l} - \sin(2\beta)\vec{m}$$

$$M_q(\vec{m}) = \sin(2\beta)\vec{l} \times \cos(2\beta)\vec{m}$$

$$M_q(\vec{n}) = \vec{n}.$$

We will establish the first of these and leave the other two as an exercise.

Applying Proposition 3.1 to vectors gives, for example,

$$\vec{n}\vec{l} = V(\vec{n}\vec{l}) + S(\vec{n}\vec{l})$$

$$= \vec{n} \times \vec{l} - \vec{n} \cdot \vec{l} = \vec{m} - 0.$$

Thus

$$M_q(\vec{l}) = (\cos\beta - \vec{n}\sin\beta)\,\vec{l}(\cos\beta + \vec{n}\sin\beta)$$

$$= (\cos\beta\vec{l} - \sin\beta\,\{\vec{n} \times \vec{l} - \vec{n} \cdot \vec{l}\}\,)\,(\cos\beta + \vec{n}\sin\beta)$$

$$= \cos^2\beta\vec{l} - \sin\beta\cos\beta\,\vec{m} + \cos\beta\sin\beta\,\{\vec{l} \times n - \vec{l} \cdot \vec{n}\}$$

$$- \sin^2\beta\,\{\vec{m} \times \vec{n} - \vec{m} \cdot \vec{n}\}$$

$$= (\cos^2\beta - \sin^2\beta)\,\vec{l} - 2\sin\beta\cos\beta\,\vec{m}$$

$$= \cos(2\beta)\,\vec{l} - \sin(2\beta)\,\vec{m}.$$

The proof is completed in Exercise 4.3.

Recall that the axis/angle representation was not unique. The pair $\{\vec{n}, \Phi\}$ and the pair $\{-\vec{n}, 2\pi - \Phi\}$ both describe the same rotation. This yields

$$\cos(\Phi/2) + \vec{n}\sin(\Phi/2) = \mathbf{q}$$

and

$$\cos\left(\frac{2\pi - \Phi}{2}\right) - \vec{n}\sin\left(\frac{2\pi - \Phi}{2}\right)$$

$$= -\cos\left(\frac{\Phi}{2}\right) - \vec{n}\sin\left(\frac{\Phi}{2}\right) = -\mathbf{q}.$$

Thus **q** and $-\mathbf{q}$ represent the same rotation and give the same attitude matrix A.

Finally we note how easy it is to compose two rotations (or equivalently to multiply by successive attitude matrices), and to express the result as a new rotation using quaternions.

Recall

$$M_{\mathbf{p}} \circ M_{\mathbf{q}} = M_{\mathbf{qp}}$$

or

$$M_{(\cos\gamma + \vec{r}\sin\gamma)} \circ M_{(\cos\beta + \vec{n}\sin\beta)}$$

$$= M_{(\{\cos\beta\cos\gamma - (\vec{n}\cdot\vec{r})\sin\beta\sin\gamma\} + \{\sin\beta\cos\gamma\,\vec{n} + \cos\beta\sin\gamma\,\vec{r}\})}.$$

The quaternion product **qp** needed to find the representation for the composition involves fewer algebraic operations than the matrix multiplication of successive attitude matrices.

Exercises

4.1 Verify that $\{M_{\mathbf{q}}(\mathbf{p})\}^* = M_{\mathbf{q}}(\mathbf{p}^*)$.

4.2. Generalize Theorem 4.2 by showing that the scalar part of $M_{\mathbf{q}}(\mathbf{p})$ equals the scalar part of $\mathbf{p}$, for any $\mathbf{p} \in H$.

4.3. Verify the two identities above which are needed to complete the proof of Theorem 4.3.

4.4. Find the three quaternions which represent $A_3(\psi)$, $A_1(\phi)$ and $A_2(\theta)$, respectively. Then find the quaternion representing $A_{3\text{-}1\text{-}2}$ by multiplying the three quaternions in the appropriate order.

4.5. Find the quaternions representing the attitude matrix in Exercise 2.4.

4.6. Find the attitude matrix A, the rotation axis $\vec{n}$, and the rotation angle Φ corresponding to the unit quaternion

$$\mathbf{q} = \frac{1}{\sqrt{15}}(2\mathbf{i} - \mathbf{j} + \mathbf{k} + 3).$$

5. Kinematic Equations for Attitude

5.1 Dynamics and Kinematics

The study of motion as the result of physical forces is called *dynamics*. NASA scientists have found it impossible to come up with an accurate dynamical model for satellite attitude motion. There are simply too many different small forces causing rotational motion, and each is difficult to model. For example, rotational torques are caused by the internal motion of tape recorders and scientific instruments, atmospheric drag, and the variations in solar pressure, magnetic attraction, and gravity gradients between different parts of the spacecraft. These torques are not ignored. In fact every one of these forces is sometimes deliberately used to control attitude motion. However, the onboard control systems are almost always based upon the kinematic equations of attitude.

Kinematics is the study of motion irrespective of the forces which bring about that motion. The kinematic equations which we will derive express the rate of change of attitude with respect to time in terms of the instantaneous attitude and angular velocity vector. Horizon scanners, sun sensors, and star trackers are used in various satellites to measure attitude. Onboard gyroscopes are used to obtain estimates of the angular velocity vector.

The attitude of a satellite is not fixed but rather is constantly changing with time. Thus all of the parametrizations which we have studied are functions of time, and we now consider the time dependence of $A(t)$, $n(t)$, $\Phi(t)$, $\psi(t)$, $\phi(t)$, $\theta(t)$, and $\mathbf{q}(t)$. We note that the derivative of a matrix function, like the derivative of a vector function, is the derivative of each component.

5.2 Quaternion Kinematics

Consider $\mathbf{q}(t_0)$ and $\mathbf{q}(t_0 + \triangle t)$ representing rotations from the reference frame to the body system at the time t_0 and $(t_0 + \triangle t)$, respectively. We can find a quaternion $\triangle \mathbf{p}$ which represents the transition from the body system at time t_0 to the body system at time $(t_0 + \triangle t)$. Then $\mathbf{q}(t_0 + \triangle t) = \mathbf{q}(t_0)\triangle \mathbf{p}$.

Suppose that $\triangle \vec{n}$ is the axis of rotation and that $\triangle \Phi$ is the angle of rotation for $\triangle \mathbf{p}$ (where for definiteness $\triangle \Phi$ is small in absolute value and has the same sign as $\triangle t$). Measured in the reference frame,

$$\triangle\vec{n} = n_u\,\vec{u}(t_0) + n_v\,\vec{v}(t_0) + n_w\,\vec{w}(t_0)$$

where $\{\vec{u}(t_0), \vec{v}(t_0), \vec{w}(t_0)\}$ is the right-hand triad representing the body system at time t_0. However, for the purposes of constructing $\triangle\mathbf{p}$, we look from the body system and use

$$\triangle\vec{n}_b = n_u\vec{i} + n_v\vec{j} + n_w\vec{k}.$$

Thus

$$\triangle\mathbf{p} = \cos(\triangle\Phi/2) + \triangle\vec{n}_b \sin(\triangle\Phi/2).$$

The instantaneous *angular velocity vector* is

$$\vec{\omega}(t_0) = \lim_{\triangle t \to 0} \frac{\triangle\Phi\,\triangle\vec{n}}{\triangle t}$$

$$= \omega_u\,\vec{u}(t_0) + \omega_v\,\vec{v}(t_0) + \omega_w\,\vec{w}(t_0).$$

For some of the algebra that follows, it is convenient to use

$$\vec{\omega}_b\,(t_0) = \omega_u\vec{i} + \omega_v\vec{j} + \omega_w\vec{k}.$$

We note that

$$\vec{\omega}_b(t_0) = \lim_{\triangle t \to 0} ((\triangle\Phi\,\triangle\vec{n}_b)/\triangle t).$$

Using Proposition 3.2 and then decomposing the matrix,

$$\mathbf{q}(t_0 + \triangle t) = \left[\cos\,(\frac{\triangle\Phi}{2})\,I_4 + \sin\,(\frac{\triangle\Phi}{2})\,\widetilde{Q}(\triangle n_b)\right]\,\mathbf{q}(t_0).$$

We can now find the derivative of $\mathbf{q}$ at t_0 by the following:

$$\frac{d\mathbf{q}}{dt}\,(t_0) = \lim_{\triangle t \to 0} \frac{\mathbf{q}(t_0 + \triangle t) - \mathbf{q}(t_0)}{\triangle t}$$

$$= \lim_{\triangle t \to 0} \frac{1}{\triangle t} \left[\left[\cos\,(\frac{\triangle\Phi}{2}) - 1\right] I_4 + \sin\,(\frac{\triangle\Phi}{2})\,\widetilde{Q}(\triangle n_b)\right]\,\mathbf{q}(t_0)$$

$$= \lim_{\Delta t \to 0} \left[\left[\frac{\cos(\frac{\Delta\Phi}{2}) - 1}{(\Delta\Phi/2)} \right] \frac{(\frac{\Delta\Phi}{2})}{\Delta t} I_4 + \frac{1}{2} \frac{\sin(\frac{\Delta\Phi}{2})}{(\Delta\Phi/2)} \tilde{Q} \left[\frac{\Delta\Phi \, \Delta\vec{n}_b}{\Delta t} \right] \right] \mathbf{q}(t_0)$$

$$= \frac{1}{2} \tilde{Q}(\vec{\omega}_b(t_0)) \, \mathbf{q}(t_0).$$

Using Proposition 3.2 in the opposite way, this can be rewritten as

$$\frac{d\mathbf{q}}{dt}(t_0) = \frac{1}{2} \mathbf{q}(t_0) \, \vec{\omega}_b(t_0).$$

Finally this can be written out in components

$$\begin{bmatrix} \dot{q}_1 \\ \dot{q}_2 \\ \dot{q}_3 \\ \dot{q}_4 \end{bmatrix} = \frac{1}{2} \begin{bmatrix} 0 & \omega_w & -\omega_v & \omega_u \\ -\omega_w & 0 & \omega_u & \omega_v \\ \omega_v & -\omega_u & 0 & \omega_w \\ -\omega_u & -\omega_v & -\omega_w & 0 \end{bmatrix} \begin{bmatrix} q_1 \\ q_2 \\ q_3 \\ q_4 \end{bmatrix}.$$

Since the components of the angular velocity vector can be estimated from onboard gyroscope readings, the predicted attitude can be computed by numerically solving this system of linear differential equations. Further, these equations can be used in the design of the automatic control system so that corrective action can be taken by small jets, momentum wheels, or electromagnets to maintain a desired orientation. The most important point to note is that these kinematic equations are linear and everywhere defined.

5.3 Attitude Matrix Kinematics

Similarly,

$$A(t_0 + \Delta t) = (\Delta A) \, A(t_0)$$

$$\Delta A = \cos\Delta\Phi \, I_3 + (1 - \cos\Delta\Phi) \begin{bmatrix} n_u \\ n_v \\ n_w \end{bmatrix} (n_u, n_v, n_w) + \sin\Delta\Phi \, Q(\Delta n_b).$$

$$\frac{dA}{dt}(t_0) = \lim_{\Delta t \to 0} \frac{A(t_0 + \Delta t) - A(t_0)}{\Delta t}$$

$$= \lim_{\Delta t \to 0} \left[\left[\frac{\cos \Delta\Phi - 1}{\Delta\Phi} \frac{\Delta\Phi}{\Delta t} \right] \left[I_3 - \begin{bmatrix} n_u \\ n_v \\ n_w \end{bmatrix} (n_u, n_v, n_w) \right] + \frac{\sin \Delta\Phi}{\Delta\Phi} Q \left[\frac{\Delta\Phi \, \Delta\vec{n}_b}{\Delta t} \right] A(t_0) \right]$$

$$= Q(\vec{\omega}_b(t_0)) \; A(t_0).$$

Written out,

$$\begin{bmatrix} \dot{u}_1 & \dot{u}_2 & \dot{u}_3 \\ \dot{v}_1 & \dot{v}_2 & \dot{v}_3 \\ \dot{w}_1 & \dot{w}_2 & \dot{w}_3 \end{bmatrix} = \begin{bmatrix} 0 & \omega_w & -\omega_v \\ -\omega_w & 0 & \omega_u \\ \omega_v & -\omega_u & 0 \end{bmatrix} \begin{bmatrix} u_1 & u_2 & u_3 \\ v_1 & v_2 & v_3 \\ w_1 & w_2 & w_3 \end{bmatrix}.$$

5.4 Euler Angle Kinematics

Consider the 3-1-2 sequence $\{\psi, \phi, \theta\}$, and let

$$q(t) = \mathbf{s}(\psi(t)) \; \mathbf{p}(\phi(t)) \; \mathbf{r}(\theta(t))$$

where $\mathbf{s} = \cos(\psi/2) + \mathbf{k} \sin(\psi/2)$, $\mathbf{p} = \cos(\phi/2) + \mathbf{i} \sin(\phi/2)$, and $\mathbf{r} = \cos(\theta/2) + \mathbf{j} \sin(\theta/2)$. The "product rule" holds for differentiating a quaternion product.

$$\frac{dq}{dt} = \left(\frac{d\mathbf{s}}{d\psi} \frac{d\psi}{dt}\right) \mathbf{p} \, \mathbf{r} + \mathbf{s} \left(\frac{d\mathbf{p}}{d\phi} \frac{d\phi}{dt}\right) \mathbf{r} + \mathbf{s} \, \mathbf{p} \left(\frac{d\mathbf{r}}{d\theta} \frac{d\theta}{dt}\right).$$

The left-hand side we have found to be

$$\frac{1}{2} \mathbf{q} \vec{\omega}_b = \frac{1}{2} \mathbf{s} \, \mathbf{p} \, \mathbf{r} \, \vec{\omega}_b$$

Solving for ω_b we have

$$\vec{\omega}_b = 2\mathbf{r}^{-1}\mathbf{p}^{-1}\mathbf{s}^{-1}\left(\frac{d\mathbf{s}}{d\psi^{\psi}}\right)\mathbf{p}\,\mathbf{r} + \mathbf{s}\left(\frac{dp}{d\phi^{\phi}}\right)\mathbf{r} + \mathbf{s}\,\mathbf{p}\left(\frac{d\mathbf{r}}{d\theta^{\theta}}\right) \ .$$

Although it is somewhat tedious to multiply all of this out and perform the indicated differentiations, the process is straightforward and yields (in components)

$$\omega_u = -\sin\theta\cos\phi\,\dot{\psi} + \cos\theta\,\dot{\phi}$$

$$\omega_v = \sin\phi\,\dot{\psi} + \dot{\theta}$$

$$\omega_w = \cos\theta\cos\phi\dot{\psi} + \sin\theta\,\dot{\phi}.$$

Solving for the derivatives of the Euler angles,

$$\dot{\psi} = (\omega_w\cos\theta - \omega_u\sin\theta)\sec\phi$$

$$\dot{\phi} = \omega_u\cos\theta + \omega_w\sin\theta$$

$$\dot{\theta} = \omega_v + (\omega_u\sin\theta - \omega_w\cos\theta)\tan\phi.$$

Notice that the kinematic equations for attitude in terms of this Euler angle sequence are nonlinear. Further they are undefined when $\phi = \pi/2$ or $3\pi/2$.

Exercises

5.1. Verify the product rule for the quaternion product by showing that

$$\frac{d}{dt}(\mathbf{q}(t)\ \mathbf{p}(t)) = \frac{d\mathbf{q}(t)}{dt}\mathbf{p}(t) + \mathbf{q}(t)\frac{d\mathbf{p}(t)}{dt}.$$

5.2. An alternative method for finding dA/dt is given below. Verify the steps and complete the process.

$$\begin{bmatrix} \frac{dA}{dt} & 0 \\ 0 & 0 \end{bmatrix} = \frac{d\widetilde{A}}{dt} = \left[\frac{d}{dt}M_{\mathbf{q}}(\mathbf{i}),\ \frac{d}{dt}M_{\mathbf{q}}(\mathbf{j}),\ \frac{d}{dt}M_{\mathbf{q}}(\mathbf{k}),\ \frac{d}{dt}M_{\mathbf{q}}(\mathbf{1})\right].$$

Find $(d/dt)(M_{\mathbf{q}}(\mathbf{p}))$ for $\mathbf{p} = \mathbf{i}, \mathbf{j}, \mathbf{k}$, and $\mathbf{1}$.

5.3. In a manner similar to that used in Section 5.4 for Euler angles, find the kinematic equations for the axis of rotation $n(t)$ and the angle of rotation $\Phi(t)$, but do not solve for $\vec{n}$ and Φ.

5.4. (For students with a knowledge of systems of differential equations.) Suppose $\mathbf{q}(0) = 1/\sqrt{2} + \mathbf{i}(1/\sqrt{2})$ and that $\vec{\omega}(t) = 2$ radian/sec $\vec{i} + 4$ radian/sec $\vec{k}$ for all $t \geq 0$. Solve the kinematic equations.

6. Bibliography

H. Anton, *Elementary Linear Algebra,* 3rd, Wiley & Sons, New York, 1981.

E. T. Bell, *Men of Mathematics,* Simon and Schuster, New York, 1937 (Chapter 19).

H. S. M. Coxeter, *Regular Complex Polytopes,* Cambridge University Press, London, 1974.

M. J. Crowe, *A History of Vector Analysis,* University of Notre Dame Press, Notre Dame, IN, 1967.

H. Goldstein, *Classical Mechanics,* Addison-Wesley, Reading, MA, 1950.

Sir W. R. Hamilton, *Elements of Quaternions,* Longmans, Green and Co., London, 1866.

J. R. Wertz, ed., *Spacecraft Attitude Determination and Control,* Reidel, Holland, 1978.

7. Acknowledgement

The author wishes to thank Professor Heath K. Riggs of the University of Vermont, who introduced him to some of the mathematics of space flight. This interest was greatly nurtured by a 1982 NASA-ASEE Summer Faculty Fellowship. Further thanks go to all of the members of the former Attitude Dynamics Section, headed by Mr. Eugene J. Lefferts and located in the Mission and Data Operations Directorate of the Goddard Space Flight Center, Greenbelt, Maryland.

8. Model Examination

Given the length and complexity of the formulas relating the various representations for attitude, it is assumed that the text of this module is available to anyone working this model examination.

1. Verify the claim (Section 2.5) that any 3 × 3 matrix B satisfying

$$BB^{T} = I, \qquad \det B = 1,$$

and $$B \begin{bmatrix} 0 \\ 0 \\ 1 \end{bmatrix} = \begin{bmatrix} 0 \\ 0 \\ 1 \end{bmatrix}$$

must be of the form

$$\begin{bmatrix} \cos\Phi & \sin\Phi & 0 \\ -\sin\Phi & \cos\Phi & 0 \\ 0 & 0 & 1 \end{bmatrix}.$$

2. Carry out the indicated operations in quaternion algebra.

(a) $(\mathbf{i} - 2\mathbf{j} + 7\mathbf{k} - 3) + (3\mathbf{i} + 4\mathbf{j} - 6)$.

(b) $(\mathbf{i} + \mathbf{j} + \mathbf{k} + 2)(3\mathbf{i} - 5\mathbf{k})$.

(c) $|2\mathbf{i} - 5\mathbf{j} + 1|$.

(d) $\mathbf{q} = 2\mathbf{i} + \mathbf{j} - \mathbf{k} + 3$, find $\mathbf{q}^{-1}$.

3. Convert each of the following to a quaternion representation.

(a) $$A = \begin{bmatrix} \frac{1}{\sqrt{2}} & 0 & \frac{1}{\sqrt{2}} \\ \frac{1}{\sqrt{3}} & \frac{1}{\sqrt{3}} & -\frac{1}{\sqrt{3}} \\ -\frac{1}{\sqrt{6}} & \frac{2}{\sqrt{6}} & \frac{1}{\sqrt{6}} \end{bmatrix} \quad \text{(attitude matrix).}$$

(b) $\vec{n} = \frac{1}{\sqrt{3}}(\vec{i} + \vec{j} + \vec{k}),\ \Phi = \frac{\pi}{2}$

(angle and axis of rotation).

(c) $\psi = \pi/2,\ \phi = \pi/2,\ \theta = \pi$ (3-1-2 Euler angles).

4. Find the following alternative representations for the attitude represented by the quaternion with

$$q_1 = 0.1392282829$$

$$q_2 = 0.0464094276$$

$$q_3 = -0.0928188552$$

$$q_4 = 0.984807753.$$

(a) The axis and angle of rotation, $\vec{n}$ and Φ.

(b) The attitude matrix A.

(c) (Optional) The 3-1-2 Euler angles ψ, ϕ, and θ.

5. For any function f of three variables, we have the following generalized Taylor approximation.

$$f(x, y, z) \sim f(x_0, y_0, z_0) + (\Delta x, \Delta y, \Delta z)\nabla f(x_0, y_0, z_0)$$

(LINEAR)

$$\sim f(x_0, y_0, z_0) + (\Delta x, \Delta y, \Delta z)\nabla f(x_0, y_0, z_0)$$
$$+ \frac{1}{2}(\Delta x, \Delta y, \Delta z)\, J(f; x_0, y_0, z_0)\, (\Delta x, \Delta y, \Delta z)^t$$

(QUADRATIC)

where $(\Delta x, \Delta y, \Delta x) = (x-x_0, y-y_0, z-z_0)$, $\nabla f = (f_x, f_y, f_z)^t$ and

$$J(f) = \begin{bmatrix} f_{xx} & f_{xy} & f_{xz} \\ f_{yx} & f_{yy} & f_{yz} \\ f_{zx} & f_{zy} & f_{zz} \end{bmatrix}$$

Given a 3-1-2 Euler Sequence ψ, ϕ, θ, we can consider *each component* of the corresponding unit quaternion $\mathbf{q}$ as a function of ψ, ϕ, and θ by

$$\mathbf{q} = \{\sin \frac{\psi}{2} \mathbf{k} + \cos \frac{\psi}{2}\} \{\sin \frac{\phi}{2} \mathbf{i} + \cos \frac{\phi}{2}\} \{\sin \frac{\theta}{2} \mathbf{j} + \cos \frac{\theta}{2}\}.$$

Use the linear Taylor approximation about $(\psi, \phi, \theta) = (0, 0, 0)$ for each *component* of $\mathbf{q}$ to justify the claim that for small angles

$$q_1 \simeq \frac{\phi}{2}, \ q_2 \sim \frac{\theta}{2}, \ q_3 \simeq \frac{\psi}{2}, \ q_4 \simeq 1.$$

Find the quadratic approximations as well.

9. Answers to Exercises

Exercises

2.1. $\vec{w} = \vec{u} \times \vec{v} = (2, -2, \sqrt{2})/\sqrt{10}$.

$$A = \frac{1}{\sqrt{10}} \begin{bmatrix} \sqrt{5} & \sqrt{5} & 0 \\ -1 & 1 & 2\sqrt{2} \\ 2 & -2 & \sqrt{2} \end{bmatrix}.$$

2.2 For example,

$$A = \begin{bmatrix} 1 & 0 & 0 \\ 0 & 1 & 0 \\ 0 & 0 & -1 \end{bmatrix}.$$

Note that any orthogonal matrix A preserves the dot product in the sense that for any $\vec{a}$ and $\vec{b}$

$$(A\vec{a}) \cdot (A\vec{b}) = \vec{a}^t A^t A \vec{b} = \vec{a}^t b = \vec{a} \cdot \vec{b}.$$

Thus if $\{\vec{u}, \vec{v}, \vec{w}\}$ is a right-hand triad, then $\{A\vec{u}, A\vec{v}, A\vec{w}\}$ is a set of orthonormal vectors. We need only check orientation. Using the matrix A above,

$$(A\vec{u})\ (A\vec{v}) = (-u_2v_3 + u_3v_2)\vec{i} + (u_1v_3 - u_3v_1)\vec{j}$$
$$+ (u_1v_2 - u_2v_1)\vec{k}.$$

But

$$\vec{w} = \vec{u} \times \vec{v}$$
$$= (u_2v_3 - u_2v_3)\vec{i} + (-u_1v_3 + u_3v_1)\vec{j}$$
$$+ (u_1v_2 - u_2v_1)\vec{k}.$$

Thus

$$(A\vec{u}) \times (A\vec{v}) = -(A\vec{w}) = -\{(A\vec{v}) \times (A\vec{u})\}.$$

2.3. Case 1. $n_1{}^2 + n_2{}^2 \neq 0$. Let $\vec{l} = (n_2\vec{i} - n_1\vec{j})/N$, where

$$N = \sqrt{n_1{}^2 + n_2{}^2},$$

and

$$\vec{m} = \vec{n} \times \vec{l} = (n_1n_3\vec{i} + n_2n_3\vec{j} - N^2\vec{k})/N.$$

Let

$$P = \begin{bmatrix} l_1 & l_2 & l_3 \\ m_1 & m_2 & m_3 \\ n_1 & n_2 & n_3 \end{bmatrix},$$

the transition matrix from $\{\vec{i}, \vec{j}, \vec{k}\}$ to $\{\vec{l}, \vec{m}, \vec{n}\}$, and

$$B = \begin{bmatrix} \cos\phi & \sin\phi & 0 \\ -\sin\phi & \cos\phi & 0 \\ 0 & 0 & 1 \end{bmatrix}.$$

Finally compute

$$A = P^t B P$$

$$= \cos\Phi\, P^t \begin{bmatrix} 1 & 0 & 0 \\ 0 & 1 & 0 \\ 0 & 0 & 0 \end{bmatrix} P + \sin\Phi\, P^t \begin{bmatrix} 0 & 1 & 0 \\ -1 & 0 & 0 \\ 0 & 0 & 0 \end{bmatrix} P$$

$$+ P^t \begin{bmatrix} 0 & 0 & 0 \\ 0 & 0 & 0 \\ 0 & 0 & 1 \end{bmatrix} P.$$

2.4. Find $\vec{n}$ as a unit eigenvector for $\lambda = 1$ and the matrix $A = A_1(\phi)\, A_3(\psi)$. Then use the expression for A in Exercise 2.3 to find cos Φ and sin Φ. For example,

$$\vec{n} = (\sqrt{2} + 1, 1, \sqrt{3}) / \sqrt{7 + 2\sqrt{2}},$$

$$\Phi = \cos^{-1}\left(\frac{3\sqrt{2} - 2}{8}\right) \sim 73.72°.$$

2.5. This fact is easily visualized geometrically. Or use Exercise 2.3, noting that we get the same attitude matrix because

$$\cos(2\pi - \Phi) = \cos\Phi, \quad \sin(2\pi - \Phi) = -\sin\Phi$$

and the expressions involving $(-\vec{n})$ and $\vec{n}$ take the appropriate sign.

3.1. **(a)** $\mathbf{i} - 2\mathbf{j} + 8\mathbf{k} - 2$.

(b) $4\mathbf{i} - 5\mathbf{j} + 8\mathbf{k} - 15$.

(c) $6\mathbf{i} - 30\mathbf{j} + 12\mathbf{k} - 21$.

(d) $\sqrt{126}$.

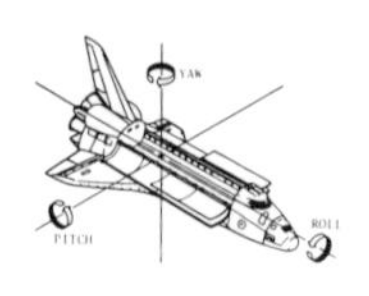

3.2. Multiplying by

$$
\begin{array}{rrrrrr}
 & & p_1\mathbf{i} + p_2\mathbf{j} & + p_3\mathbf{k} & + p_4 \\
 & & q_1\mathbf{i} + q_2\mathbf{j} & + q_3\mathbf{k} & + q_4 \\
\hline
 & & p_1q_4\mathbf{i} + p_2q_4\mathbf{j} & + p_3q_4\mathbf{k} & + p_4q_4 \\
 & -p_1q_3\mathbf{j} + p_2q_3\mathbf{i} - p_3q_3 & + p_4q_3\mathbf{k} & & \\
p_1q_2\mathbf{k} - p_2q_2 - p_3q_2\mathbf{i} + p_4q_2\mathbf{j} & & & & \\
-p_1q_1 - p_2q_1\mathbf{k} + p_3q_1\mathbf{j} + p_4q_1\mathbf{i} & & & &
\end{array}
$$

Collecting appropriate terms together gives the desired result.

3.4. $$|q_1\mathbf{i} + q_4| = \sqrt{q_1^2 + q_4^2}$$

$$(p_1\mathbf{i} + p_4)(q_1\mathbf{i} + q_4) = (p_4q_1 + q_4p_1)\mathbf{i} + (p_4q_4 - p_1q_1).$$

3.5. One way is to use Proposition 3.1.

$$
\begin{aligned}
(\mathbf{pq})^* &= (\{p_4\vec{q} + q_4\vec{p} + \vec{p} \times \vec{q}\} + \{p_4q_4 - \vec{p} \cdot \vec{q}\})^* \\
&= -p_4\vec{q} - q_4\vec{p} - (\vec{p} \times \vec{q}) + p_4q_4 - \vec{p} \cdot \vec{q}. \\
\mathbf{q}^*\mathbf{p}^* &= (-\vec{q} + q_4)(-\vec{p} + p_4) \\
&= \{q(-\vec{p}) + p_4(-\vec{q}) + (-\vec{q}) \times (-\vec{p})\} \\
&\quad + \{q_4p_4 - (-\vec{q}) \cdot (-\vec{p})\} \\
&= -q_4\vec{p} - p_4\vec{q} - (\vec{p} \times \vec{q}) + q_4p_4 - \vec{p} \cdot \vec{q}.
\end{aligned}
$$

3.6. Suppose that $\mathbf{q}^2 = -1$.

$$\{q_4\vec{q} + q_4\vec{q} + \vec{q} \times \vec{q}\} + \{q_4^2 - \vec{q} \cdot \vec{q}\} = -1.$$

Since $\vec{q} \times \vec{q} = \vec{0}$, this implies that $2q_4\vec{q} = \vec{0}$. If $\vec{q} = \vec{0}$, then we must have $q_4^2 = -1$, which is impossible. Thus $q_4 = 0$ and $-\vec{q} \cdot \vec{q} = -1$. Finally

$$|\mathbf{q}| = \sqrt{q_1^2 + q_2^2 + q_3^2 + 0^2} = \vec{q} \cdot \vec{q} = 1.$$

On the other hand, suppose $|\mathbf{q}| = 1$ and $S(\mathbf{q}) = 0$

$$\mathbf{q}^2 = \{2q_4\vec{q} + \vec{q} \times \vec{q}\} + \{q_4^2 - \vec{q} \cdot \vec{q}\} = -\vec{q} \cdot \vec{q} = -|\mathbf{q}| = -1.$$

4.1. Assume that $|\mathbf{q}| = 1$.

$$M_{\mathbf{q}}(\mathbf{p})^* = (\mathbf{q}^*\mathbf{p}\mathbf{q})^* = \mathbf{q}^*\mathbf{p}^*(\mathbf{q}^*)^*$$

$$= \mathbf{q}^*\mathbf{p}^*\mathbf{q} = M_{\mathbf{q}}(\mathbf{p}^*).$$

4.2. Assume $|\mathbf{q}| = \sqrt{q_4{}^2 + \vec{q}\ \vec{q}} = 1$.

$$\mathbf{pq} = \{p_4q_4 - \vec{p} \cdot \vec{q}\} + \{p_4\vec{q} + q_4\vec{p} + \vec{p} \times \vec{q}\}.$$

$$S(q^*\{\mathbf{pq}\}) = q_4\{p_4q_4 - \vec{p} \cdot \vec{q}\} - (-\vec{q})$$
$$\{p_4\vec{q} + q_4\vec{p} + \vec{p} \times \vec{q}\}$$

$$= p_4(q_4)^2 - q_4\vec{p} \cdot \vec{q} + p_4\vec{q} \cdot \vec{q}$$
$$+ q_4\vec{q} \cdot \vec{p} + \vec{q} \cdot (\vec{p} \times \vec{q})$$

$$= p_4(q_4{}^2 + \vec{q} \cdot \vec{q}) = p_4.$$

Note the use of properties of the scalar triple product to conclude that $\vec{q}\ \ (\vec{p} \times \vec{q}) = 0$.

4.3.

$$M_{\mathbf{q}}(\vec{m}) = (\cos\beta - \vec{n}\sin\beta)\vec{m}\,(\cos\beta + \vec{n}\sin\beta)$$

$$= (\vec{m}\cos\beta - [\vec{n} \times \vec{m} - \vec{n} \cdot \vec{m}]\sin\beta)$$
$$(\cos\beta + \vec{n}\sin\beta)$$

$$= \vec{m}\cos^2\beta + \vec{l}\sin\beta\cos\beta$$
$$+ [\vec{m} \times \vec{n} - \vec{m} \cdot \vec{n}]\sin\beta\cos\beta$$
$$+ [\vec{l} \times \vec{n} - \vec{l} \cdot \vec{n}]\sin^2\beta$$

$$= \vec{m}\,(\cos^2\beta - \sin^2\beta) + \vec{l}(2\sin\beta\cos\beta)$$

$$= \sin(2\beta)\vec{l} + \cos(2\beta)\vec{m}.$$

$$M_{\mathbf{q}}(\vec{n}) = (\cos\beta - \vec{n}\sin\beta)\vec{n}\,(\cos\beta + \vec{n}\sin\beta)$$

$$= (\cos\beta\,\vec{n} - \sin\beta\,[\vec{n} \times \vec{n} - \vec{n} \cdot \vec{n}])\,(\cos\beta + \vec{n}\sin\beta)$$

$$= \cos^2\beta\,\vec{n} + \sin\beta\cos\beta$$
$$+ \sin\beta\cos\beta\,[\vec{n} \times \vec{n} - \vec{n} \cdot \vec{n}] + \sin^2\beta\,\vec{n}$$
$$= \vec{n}.$$

4.4.

$$\mathbf{s} = \cos(\frac{1}{2}\psi) + \mathbf{k}\sin(\frac{1}{2}\psi) \text{ represents } A_3(\psi).$$

$$\mathbf{p} = \cos(\frac{1}{2}\phi) + \mathbf{i}\sin(\frac{1}{2}\phi) \text{ represents } A_1(\phi).$$

$$\mathbf{r} = \cos(\frac{1}{2}\theta) + \mathbf{j}\sin(\frac{1}{2}\theta) \text{ represents } A_2(\theta).$$

Then $A_{3\text{-}1\text{-}2} = A_2(\theta)\, A_1(\phi)\, A_3(\psi)$ is represented by

$\mathbf{q} = \mathbf{spr}$ where

$$\begin{aligned}\mathbf{q} = {} & \mathbf{i}\,(\cos\frac{1}{2}\psi \sin\frac{1}{2}\phi \cos\frac{1}{2}\theta - \sin\frac{1}{2}\psi \cos\frac{1}{2}\phi \sin\frac{1}{2}\theta) \\ & + \mathbf{j}\,(\cos\frac{1}{2}\psi \cos\frac{1}{2}\phi \sin\frac{1}{2}\theta + \sin\frac{1}{2}\psi \sin\frac{1}{2}\phi \cos\frac{1}{2}\theta) \\ & + \mathbf{k}\,(\cos\frac{1}{2}\psi \sin\frac{1}{2}\phi \sin\frac{1}{2}\theta + \sin\frac{1}{2}\psi \cos\frac{1}{2}\phi \cos\frac{1}{2}\theta) \\ & + \phantom{\mathbf{k}}\,(\cos\frac{1}{2}\psi \cos\frac{1}{2}\phi \cos\frac{1}{2}\theta - \sin\frac{1}{2}\psi \sin\frac{1}{2}\phi \sin\frac{1}{2}\theta).\end{aligned}$$

4.5. $$\mathbf{q} = (\mathbf{i}\sqrt{2+\sqrt{2}} + \mathbf{j}\sqrt{2-\sqrt{2}} + \mathbf{k}\sqrt{3}\sqrt{2-\sqrt{2}} + \sqrt{3}\sqrt{2+\sqrt{2}})/4.$$

4.6. $\vec{n} = (2, -1, 1)/\sqrt{6}$

$$\frac{1}{2}\Phi = \cos^{-1}(3/\sqrt{15}) \sim 39.23 \text{ degrees}$$

$$A = \frac{1}{15}\begin{bmatrix} 11 & 2 & 10 \\ -10 & 5 & 10 \\ -2 & -14 & 5 \end{bmatrix}.$$

5.1. We use the product rule for dot products, cross products and scalar multiplication of vector function although it could be done directly.

$$\mathbf{qp} = \{q_4 p_4 - \vec{q} \cdot \vec{p}\} + \{q_4 \vec{p} + p_4 \vec{q} + \vec{q} \times \vec{p}\}$$

$$\frac{d}{dt}[\mathbf{qp}] = \{(\dot{q}_4 p_4 + q_4 \dot{p}_4) - (\dot{\vec{q}} \cdot \vec{p} + \vec{q} \cdot \dot{\vec{p}})\}$$

$$+ \{(\dot{q}_4 \vec{p} + q_4 \dot{\vec{p}}) + (\dot{p}_4 \vec{q} + p_4 \dot{\vec{q}}) + (\dot{\vec{q}} \times \vec{p} + \vec{q} \times \dot{\vec{p}})\}$$

$$= [\{\dot{q}_4 p_4 - \dot{\vec{q}} \cdot \vec{p}\} + \{\dot{q}_4 \vec{p} + p_4 \dot{\vec{q}} + \dot{\vec{q}} \times \vec{p}\}]$$

$$+ [\{q_4 \dot{p}_4 - \vec{q} \cdot \dot{\vec{p}}\} + \{q_4 \dot{\vec{p}} + \dot{p}_4 \vec{q} + \vec{q} \times \dot{\vec{p}}\}]$$

$$= \dot{\mathbf{q}}\mathbf{p} + \mathbf{q}\dot{\mathbf{p}}.$$

5.2. This can be done directly. For example,

$$\frac{d}{dt} M_q(i) = \frac{d}{dt} \begin{bmatrix} u_1 \\ v_1 \\ w_1 \end{bmatrix} = \frac{d}{dt} \begin{bmatrix} q_1^2 - q_2^2 - q_3^2 + q_4^2 \\ 2(q_1 q_2 - q_3 q_4) \\ 2(q_1 q_3 + q_2 q_4) \end{bmatrix}$$

$$\qquad\qquad 0 \qquad\qquad\qquad 0$$

and

$$\frac{du_1}{dt} = 2q_1\dot{q}_1 - 2q_2\dot{q}_2 - 2q_3\dot{q}_3 + 2q_4\dot{q}_4$$

$$= q_1(\omega_W q_2 - \omega_v q_3 + \omega_u q_4) - q_2(-\omega_w q_1 + \omega_u q_2 + \omega_v q_4)$$

$$- q_3(\omega_v q_2 - \omega_u q_3 + \omega_w q_4) + q_2(-\omega_u q_1 - \omega_v q_3 - \omega_w q_4)$$

$$= -\omega_v(2[q_1 q_3 + q_2 q_4]) + \omega_W(2[q_1 q_2 - q_3 q_4])$$

$$= -\omega_v w_1 + \omega_w v_1.$$

Alternatively, representing quaternions as column vectors, we have

$$A = [\mathbf{q}^* \, \mathbf{iq}, \mathbf{q}^* \, \mathbf{jq}, \mathbf{q}^* \, \mathbf{kq}, 1]$$

$$\frac{dA}{dt} = [\dot{\mathbf{q}}^* \, \mathbf{iq} + \mathbf{q}^* \, \mathbf{i}\dot{\mathbf{q}}, \dot{\mathbf{q}}^* \, \mathbf{jq} + \mathbf{q}^* \, \mathbf{j}\dot{\mathbf{q}}, \dot{\mathbf{q}}^* \, \mathbf{kq} + \mathbf{q}^* \, \mathbf{k}\dot{\mathbf{q}}, 0]$$

but

$$\dot{\mathbf{q}} = \frac{1}{2}\,\mathbf{q}\vec{\omega}_b \qquad \text{and} \qquad \dot{\mathbf{q}}^* = \frac{1}{2}(-\vec{\omega}_b)\,\mathbf{q}^*.$$

Substituting and algebraically rewriting

$$\frac{dA}{dt} = \frac{1}{2}\,[-(\mathbf{q}^*\mathbf{i}\mathbf{q}\vec{\omega}_b)^* + (\mathbf{q}^*\mathbf{i}\mathbf{q}\vec{\omega}_b),$$

$$-(\mathbf{q}^*\mathbf{j}\mathbf{q}\vec{\omega}_b)^* + (\mathbf{q}^*\mathbf{j}\mathbf{q}\vec{\omega}_b),\ -(\mathbf{q}^*\mathbf{k}\mathbf{q}\vec{\omega}_b)^* + (\mathbf{q}^*\mathbf{k}\mathbf{q}\vec{\omega}_b),\ 0].$$

Each component is one-half the difference between a pure vector and its conjugate, which gives the pure vector.

$$\frac{dA}{dt} = [\mathbf{q}^*\,\mathbf{i}\,\mathbf{q}\omega_b,\ \mathbf{q}^*\,\mathbf{j}\,\mathbf{q}\omega_b,\ \mathbf{q}^*\,\mathbf{k}\,\mathbf{q}\omega_b,\ 0]$$

$$= Q(\omega_b)\ A.$$

5.3.

$$\mathbf{q}(t) = \cos\phi\,(t) + \sin\phi\,(\mathrm{t})\,\vec{n}\,(t)$$

$$\dot{\mathbf{q}}(t) = -\sin\Phi\,\dot{\Phi} + \cos\Phi\,\vec{n}\,\dot{\Phi} + \sin\Phi\,\dot{\vec{n}}$$

also

$$\dot{\mathbf{q}}(t) = \frac{1}{2}\,\mathbf{q}\vec{\omega}_b.$$

Evaluating these two lines and solving for $\vec{\omega}_b$ gives

$$\vec{\omega}_b = 2(\cos\Phi - \sin\Phi\,\vec{n})(-\sin\Phi\,\dot{\Phi} + \cos\Phi\,\vec{n}\,\dot{\Phi} + \sin\Phi\,\dot{\vec{n}}).$$

In components

$$0 = 2(\sin^2\Phi - \sin\Phi\cos\Phi)\dot{\Phi}.$$

$$\omega_u = 2[(\cos\Phi - \sin\Phi)\sin\Phi\, n_1 + \sin\Phi\cos\Phi\,\dot{n}_1 - \sin^2\Phi\,(n_2\dot{n}_3 - \dot{n}_2 n_3)].$$

$$\omega_v = 2[(\cos\Phi - \sin\Phi)\sin\Phi\, n_2 + \sin\Phi\cos\Phi\,\dot{n}_2 + \sin^2\Phi\,(n_1\dot{n}_3 - n_1\dot{n}_3)].$$

$$\omega_w = 2[(\cos\Phi - \sin\Phi)\sin\Phi\, n_3 + \sin\Phi\cos\Phi\,\dot{n}_3 - \sin^2\Phi\,(n_1\dot{n}_2 - \dot{n}_1 n_2)].$$

10. Answers to Model Exam

1. Let $B = \{b_{ij}\}$. Then the fact that $B[0\ 0\ 1]^t = [0\ 0\ 1]^t$ implies that $b_{13} = b_{23} = 0$ and $b_{33} = 1$. We can conclude many things from the various components of the matrix equation $BB^t = I$. First

$$b_{31}{}^2 + b_{32}{}^2 + 1 = 1,$$

implying that $b_{31} = b_{32} = 0$. Next

$$b_{11}{}^2 + b_{12}{}^2 = 1$$

so that there exists an angle Φ with $0 \leq \Phi < 2\pi$ so that $b_{12} = \cos\Phi$ and $b_{12} = \sin\Phi$. Finally the equations

$$b_{11}b_{21} + b_{12}b_{22} = 0$$

and

$$b_{21}{}^2 + b_{22}{}^2 = 1$$

imply that $b_{21} = -\epsilon \sin\Phi$ and $b_{22} = \epsilon \cos\Phi$ where $\epsilon = +1$ or -1. Calculating the det B, we find it equals ϵ which must then be 1.

2. (a) $4\mathbf{i} - 2j + 7\mathbf{k} - 9$.

   (b) $\mathbf{i} + 8\mathbf{j} - 13\mathbf{k} + 2$.

   (c) $\sqrt{30}$.

   (d) $\mathbf{q}^{-1} = (-2\mathbf{i} - \mathbf{j} + \mathbf{k} + 3)/15$.

3. (a) Let $A = \{a_{ij}\}$ and use the representation for A in terms of $\mathbf{q}$ from Section 4.2. For example

$$a_{11} = q_1{}^2 - q_2{}^2 - q_3{}^2 + q_4{}^2.$$

One way of solving for $\mathbf{q}$ in terms of $\{a_{ij}\}$ is the following:

$$q_4 = \frac{1}{2}(1 + a_{11} + a_{22} + a_{33})^{\frac{1}{2}} = .8204732386$$

$$q_1 = \frac{1}{4q_4}(a_{23} - a_{32}) = -.4247082003$$

$$q_2 = \frac{1}{4q_4}(a_{31} - a_{13}) = -.339851143$$

$$q_3 = \frac{1}{4q_4}(a_{12} - a_{21}) = -.1759198966.$$

The negative of each gives a second possible answer.

(b) $\mathbf{q} = \cos\frac{\pi}{4} + \sin\frac{\pi}{4}(\mathbf{i} + \mathbf{j} + \mathbf{k})/\sqrt{3}$

(c) $$\mathbf{q} = (\cos\frac{\pi}{4} + \mathbf{k}\sin\frac{\pi}{4})(\cos\frac{\pi}{4} + \mathbf{i}\sin\frac{\pi}{4})(\cos\frac{\pi}{2} + \mathbf{j}\sin\frac{\pi}{2})$$

$$= -\frac{1}{2}\mathbf{i} + \frac{1}{2}\mathbf{j} + \frac{1}{2}\mathbf{k} - \frac{1}{2}.$$

4. (a) $\Phi = 20$ degrees

$\vec{n} = (.8017837257, .2672612417, -.5345224834).$

(b) $$\begin{bmatrix} .9784616562 & -.1698944472 & -.1172547482 \\ .1957404668 & .9440002966 & .2656198458 \\ .0655627088 & -.2828415256 & .9569233064 \end{bmatrix}.$$

(c) Using c for cosine and s for sine,

$$A = \begin{bmatrix} c\psi c\phi - s\theta s\psi s\phi & c\psi s\theta + s\theta s\psi c\phi & -c\theta s\psi \\ -c\theta s\phi & c\theta c\phi & s\theta \\ s\psi s\phi + s\theta c\psi s\phi & s\psi s\phi - s\theta c\psi c\phi & c\theta c\psi \end{bmatrix}.$$

θ ref. $= \sin^{-1}(a_{23}) = 15.40325251$ degrees

$$\phi \text{ ref.} = \tan^{-1}\left[\frac{-a_{21}}{a_{22}}\right] = -11.71439782 \text{ degrees}$$

$$\psi \text{ ref.} = \tan^{-1}\left[\frac{-a_{13}}{a_{33}}\right] = 6.985804076 \text{ degrees.}$$

Checking the other entries of A, these turn out to be the appropriate selections for the inverse trigonometric functions.

5. Using the product rule and the short-hand for the cosine and sine functions

$$\frac{dq}{d\psi} = \left[\frac{1}{2} c \frac{\psi}{2} \mathbf{k} - \frac{1}{2} s \frac{\psi}{2}\right] \left[s \frac{\phi}{2} \mathbf{i} + c \frac{\phi}{2} \quad \left[s \frac{\theta}{2} \mathbf{j} + c \frac{\theta}{2}\right]\right.$$

$$\frac{d^2q}{d\psi^2} = \left[-\frac{1}{4} s \frac{\psi}{2} \mathbf{k} - \frac{1}{4} c \frac{\psi}{2}\right] \left[s \frac{\phi}{2} \mathbf{i} + c \frac{\phi}{2}\right] \left[s \frac{\theta}{2} \mathbf{j} + c \frac{\theta}{2}\right]$$

$$\frac{d^2q}{d\phi d\psi} = \left[\frac{1}{2} c \frac{\psi}{2} \mathbf{k} - \frac{1}{2} s \frac{\psi}{2}\right] \left[\frac{1}{2} c \frac{\phi}{2} \mathbf{i} - \frac{1}{2} s \frac{\phi}{2}\right] \left[s \frac{\theta}{2} \mathbf{j} + c \frac{\theta}{2}\right]$$

etc.

Evaluate first at $(\psi, \phi, \theta) = (0, 0, 0)$ and then multiply.

LINEAR

$$q_1 \approx 0 + \frac{1}{2}\phi \qquad -\frac{1}{2}\psi\,\theta$$

$$q_2 \approx 0 + \frac{1}{2}\theta \qquad +\frac{1}{2}\psi\,\phi$$

$$q_3 \approx 0 + \frac{1}{2}\psi \qquad +\frac{1}{2}\phi\,\theta$$

$$q_4 \approx 1 + 0 \qquad -\frac{1}{4}(\psi^2 + \phi^2 + \theta^2)$$

QUADRATIC APPROXIMATION

UMAP

Modules in Undergraduate Mathematics and its Applications

Module 653

The Ricker Salmon Model

Raymond N. Greenwell
and
Ho Kuen Ng

Published in cooperation with the Society for Industrial and Applied Mathematics, the Mathematical Association of America, the National Council of Teachers of Mathematics, the American Mathematical Association of Two-Year Colleges, and The Institute of Management Sciences.

COMAP

INTERMODULAR DESCRIPTION SHEET: UMAP Unit 653

TITLE: THE RICKER SALMON MODEL

AUTHORS: Raymond N. Greenwell
Department of Mathematics
Hofstra University
Hempstead, New York 11550

Ho Kuen Ng
Department of Mathematics and Computer Science
San Jose State University
San Jose, California 95192

MATH FIELD: Difference equations

APPLICATION FIELD: Ecology

TARGET AUDIENCE: Students in a differential equations or modeling course.

ABSTRACT: A difference equation model describing the dynamics of a salmon population was developed by W.E. Ricker in 1954. This unit derives the model, shows how it can be modified, and introduces the concept of maximum sustainable yield. It also shows how difference equations may lead to periodic and chaotic behavior, and a computer program enables one to explore the periods and chaos. The technique of dynamic programming is introduced to show how to maximize the income from fishing over a finite period.

PREREQUISITES: Elementary differential equations.

The UMAP Journal, Vol. V, No. 3, 1984

The Ricker Salmon Model

Raymond N. Greenwell
Department of Mathematics
Hofstra University
Hempstead, New York 11550

and

Ho Kuen Ng
Department of Mathematics
San Jose State University
San Jose, California 95192

Table of Contents

1. INTRODUCTION 1
2. THE LIFE OF A SALMON 1
3. DERIVATION OF THE MODEL 2
4. PROPERTIES OF THE MODEL 4
5. DO THE DATA REALLY FIT THE MODEL? 5
6. CHAOTIC BEHAVIOR 7
7. COMPUTER SIMULATION 11
8. DYNAMIC PROGRAMMING 13
9. OTHER MODELS 16
10. REFERENCES 17
11. ANSWERS TO EXERCISES 18

Modules and Monographs in Undergraduate Mathematics and its Applications Project (UMAP)

The goal of UMAP was to develop, through a community of users and developers, a system of instructional modules in undergraduate mathematics and its applications to be used to supplement existing courses and from which complete courses may eventually be built.

The Project was guided by a National Advisory Board of mathematicians, scientists, and educators. UMAP was funded by a grant from the National Science Foundation and is now supported by the Consortium for Mathematics and Its Applications, Inc. (COMAP), a nonprofit corporation engaged in research and development in mathematics education.

The project would like to thank all those who assisted in the production of this unit.

This material was prepared with the partial support of National Science Foundation Grant No. SPE8304192. Recommendations expressed are those of the authors and do not necessarily reflect the views of the NSF or the copyright holder.

1. Introduction

In the Pacific Northwest, the economic survival of many people hinges on the success of the salmon fisheries, which in turn depends upon the survival of the salmon (who wouldn't mind surviving, either). The fishermen would like to maximize their profits by catching as many salmon as possible, but excessive fishing could cause the salmon population to drop so low that the future of the industry would be jeopardized.

Mathematical models have been used to help determine fishing policy. If a model accurately describes the biological situation, it can provide us with information otherwise difficult to obtain, as well as confirm observations already made. Further, it can suggest new areas for biological observation.

One of the most widely used mathematical models for salmon fisheries was developed by W.E. Ricker in 1954 [11]. In this unit, we will derive the Ricker model, using a derivation simpler than that of Ricker. We will then salmon up our calculus skills to study some properties of the model and make some conclusions. Next, we will exsalmon how this rather simple model exhibits some very complex and chaotic behavior under certain circumstances. Finally, we will introduce a technique known as dynamic programming to derive additional information from the model.

2. The Life of a Salmon

We start this story in the middle of the salmon's career, when they are swimming in the ocean, growing in size and strength. After a few years of the good life, the salmon start an arduous journey upstream to their birthplace. Guided by some unknown mechanism, they swim hundreds of miles against the current, making heroic leaps over rocks and waterfalls. When they finally reach their spawning place, the female salmon lay their eggs, which are then fertilized by the male salmon. At this point, the salmon have lost a quarter of their body weight, having fasted during their long journey. They soon die in the same water in which they were born.

A few months later the eggs hatch, and the baby salmon emerge. They are vulnerable at this stage to predatory birds and fish, which would love nothing better than to gobble them up. When the survivors become large enough, they begin their journey back to the ocean, where the life cycle begins all over again, repeating the odyssey of the previous generation.

3. Derivation of the Model

Since one generation of salmon dies before the next appears, we will use a difference equation to express the population of any generation in terms of the previous one. In contrast, many population models assume a continuous change in population, and so use a differential equation.

Our model requires six assumptions. We begin with two: first, the number of eggs laid is proportional to the number of adult salmon; and second, the population of the next generation is proportional to the number of eggs laid. These assumptions seem fairly reasonable, and when we put them together we get

$$N_{t+1} \propto N_t, \tag{1}$$

where N_t is the population in year t and N_{t+1} is the population in year $t+1$. As we shall later see, this relationship is not your usual proportion, because the "constant" of proportionality varies with N_t.

If this were the whole story, the salmon population would increase exponentially; that is, we would have $N_t = N_0 k^t$, where k is the constant of proportionality.

Exercise 1. Prove this last statement.

But this unbridled growth is limited by the birds and other fish who prey upon the young salmon. Our third assumption is that, until they reach a certain size, salmon are eaten at rates proportional to their number. If we let R be the population of the new generation, known as recruits, then

$$\frac{dR}{dt} = -cR, \tag{2}$$

where c is a constant or proportionality. Notice that we use a differential equation here, since R is large and the predation is going on continuously over a period of time. Solving Eq. (2), we get

$$R = R_0 e^{-ct}, \tag{3}$$

where R_0 is the initial recruit population.

Exercise 2. Derive Eq. (3).

Our fourth assumption is that after a time T, the young salmon become too big for most predators to swallow, and so their population stops decreasing. Our fifth assumption is that T is proportional to the number of eggs laid, which we already assumed to be proportioned to N_t, the adult population. The rationale here is that if there are twice as many baby salmon, and they have the same amount of food to go around, it will take twice as long for them to reach that critical size at which they can no longer be eaten easily. This assumption may seem less plausible than the others, but it is not too far-fetched, and a more reasonable assumption may make our model too complex to analyze. So we will assume

$$T = KN_t \,, \tag{4}$$

where K is another constant of proportionality. Putting T in for t in Eq. (3) yields

$$R = R_0 \, e^{-CKN_t} . \tag{5}$$

Finally, we assume that the number of adults in the next generation is proportional to the number of recruits, as given in Eq. (5). Putting this together with Eq. (1), we have

$$N_{t+1} \propto N_t \, e^{-CKN_t} , \tag{6}$$

since, for N_{t+1} to be proportional to two quantities, it must be proportional to their product. We will let the constant of proportionality be e^r, so that (6) can be rewritten as

$$N_{t+1} = N_t e^r \, e^{-CKN_t} = N_t \, e^{r(1-(CK/r)N_t)} . \tag{7}$$

Notice that if $N_t = r/CK$, then $N_{t+1} = N_t$, and hence all subsequent populations also equal r/CK. This is a special value of the population known as the *equilibrium population,* and we will denote this value by P. If the population ever exactly equals P, our model predicts that it will stay there forever. We will assume that r is positive; otherwise the equilibrium population will not exist, and, in fact, the population will get smaller as time passes. We can then simplify (7) as

$$N_{t+1} = N_t \, e^{r(1-N_t/P)} . \tag{8}$$

This is the form of the Ricker model we will usually work with.

Sometimes, rather than measuring the population N_t directly, we will look at the fraction of the equilibrium population by denoting $X_t = N_t/P$. Then, when the population is at equilibrium, $X_t = 1$. Eq. (8) becomes

$$X_{t+1} = X_t\, e^{r(1-X_t)}. \tag{9}$$

Since our model is based on six assumptions, it is only as valid as those assumptions; a model is only a model and should not be confused with the real thing. Nevertheless, it is a first step toward a quantitative understanding of salmon population. When treated with caution, the results can be helpful.

4. Properties of the Model

Let us see what we can learn from the model. Denote $N_{t+1} = f(N_t)$, where

$$f(N) = N\, e^{r(1-N/P)}.$$

Exercises 3. Show that $f'(N) = (1 - r\, N/P)\, e^{r(1-N/P)}$.

4. Show that $\lim_{N \to \infty} f(N) = 0$.

Looking at the result of Exercise 3, we see that when $N < P/r$, $f'(N) > 0$, so f is increasing. Also, f is decreasing when $N > P/r$. Thus, the population of the next generation is greatest when $N = P/r$. This value of N is known as the *maximum recruitment level*. At this point we have $f(P/r) = (P/r)\, e^{r-1}$. Coupling this fact with the result of Exercise 4 and the fact that $f(0) = 0$, we can graph f roughly as shown in Fig. 1.

In Fig. 1 we have also drawn a 45° line. Where it crosses the graph of f, $N = f(N) = P$. We have also drawn P to the right of P/r, indicating $r > 1$. It is also possible to have $r < 1$ and P/r to the right of P. The salmon population will grow if $f(N) > N$ (the part of the curve above the 45° line). Or we can catch all of the surplus population, $f(N) - N$, and be left with a population identical in size to the last generation. The fishing industry is interested in the population that will give it the maximum harvest of this type, so it can continue to harvest this same maximum harvest year after year. This is known as the level of *maximum sustainable yield.*

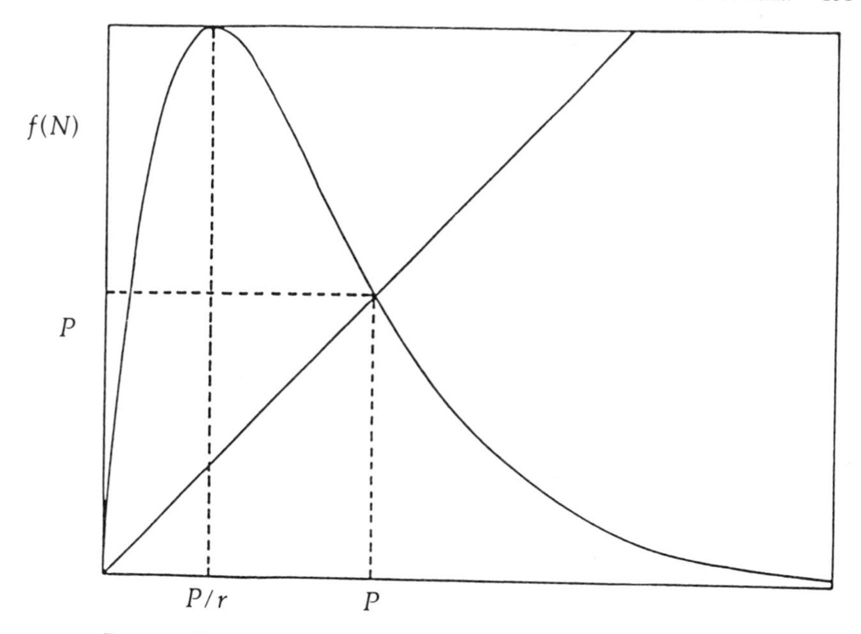

Figure 1. The graph of $F(N)$ (population of the next generation) vs. N (population of this generation).

Exercise 5. Show that the maximum sustainable yield occurs when $f'(N) = 1$.

We can call the value found in the last exercise N^*. We cannot determine N^* analytically, but we can use a numerical scheme such as Newton's method if we already have values for r and P. Once we know N^*, it is not hard to find the harvest.

Exercise 6. Show that the maximum sustainable yield is

$$N^*\left(\frac{1}{1-rN^*/P} - 1\right).$$

5. Do the Data Really Fit the Model?

That's a good question. Fig. 2 shows some of the data from [12].

We must admit, albeit with some disappointment, that we could fit almost any curve with equal success (or lack thereof) through this motley group of points in Fig. 2.

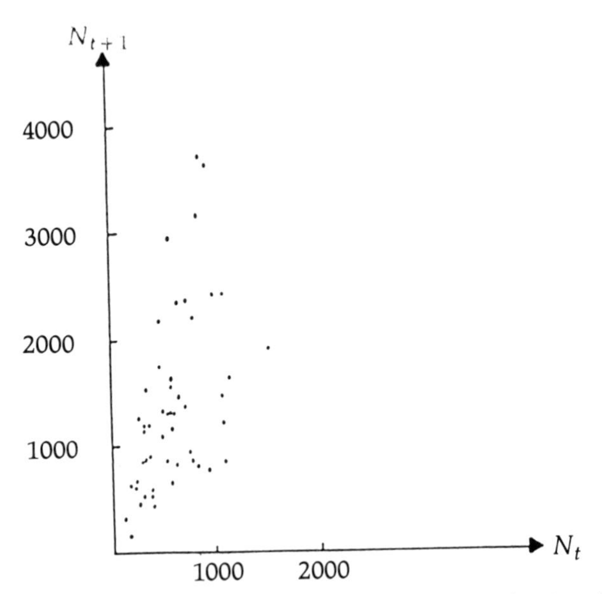

Figure 2. The plot of N_{t+1} vs. N_t for data taken from [12].

Part of the problem is that our model is deterministic: the size of each generation is completely determined by the size of the previous generation. In reality, there are other factors, such as climate and food supply, which, as far as our model is concerned, are random variables. Some have suggested that a random factor be added to the model:

$$N_{t+1} = N_t \, e^{r(1-N_t/P) \, + \, \sigma_t} \, , \tag{10}$$

where σ_t is a normal random variable.

Another problem is that the model may need modification. Thomas et al. [13] found that data for fruit flies (which are not closely related to the salmon, although the salmon might make a nice snack out of them), fit the $\theta -$Ricker model:

$$N_{t+1} = N_t \, e^{r(1-(N_t/P)^{\theta})} . \tag{11}$$

If $\theta = 1$, Eq. (11) reduces to the regular Ricker model in Eq. 8. It can be derived in a manner analogous to the derivation of the regular Ricker model by assuming that the time for the recruits to reach a less vulnerable size is proportional to the number of eggs laid, raised to the θ power. This is highly reasonable if $\theta < 1$, for it says that if the number of eggs laid is doubled, the rate that the young fruit flies (or salmon) grow should not be reduced by a full factor of two, due to a saturation effect. In the data from Thomas et al., θ is less than 1 in 52 out of 58 times. But then, these are not salmon anyway, and before using this model to study fruit flies, we should see whether our original six assumptions are valid for fruit flies. (Ricker's original derivation used a different set of assumptions to arrive at the same model.)

The process of modifying a model, checking it against the data, and then modifying it again is typical in mathematical modeling, and the cycle can go on indefinitely. Each time through the cycle the model usually becomes more complex and difficult to analyze, while doing a better job of reflecting reality. But we will stop the cycle right now by staying with the original Ricker model through the rest of this unit.

A commonly used procedure to find r and P, when presented with a series of populations $N_0, N_1, \ldots, N_n$, is to introduce a new variable $y_t = ln(N_{t+1}/N_t)$ and plot y_t vs. 5_t. If the original data fit the Ricker model (Eq. 8), then the plot of points (N_t, y_t) should lie along a straight line.

Exercise 7. Find the equation of the line relating y_t to N_t.

A technique known as the method of least squares can then estimate r and P. For more information on this method, see [1].

6. Chaotic Behavior

Suppose the salmon population at some time is close to, but not equal to, the equilibrium population P. It would be nice if the population gradually got closer to P, or at least didn't slip any further away. If, instead, the population does not approach its equilibrium value, we would like to know what it does instead. Depending on the value of r, the population may fluctuate randomly, with no apparent pattern. Furthermore, it turns out that this is true for any model $N_{t+1} = f(N_t)$ whose graph f has a hump in it, as in Fig. 1.

For simplicity, let us use Eq. (9) rather than Eq. (8) for our model. Denote $X_{t+1} = F(X_t)$. Then the equilibrium point is $P = 1$. The equilibrium population P is said to be *locally stable* if, when a value X_t is close to P, the next value X_{t+1} is no further away.

Mathematically, this says

$$|X_{t+1} - P| \leq |X_t - P| \tag{12}$$

or

$$\frac{|X_{t+1} - P|}{|X_t - P|} \leq 1.$$

But $X_{t+1} = F(X_t)$ and $P = F(P)$, so we have

$$\frac{F(X_t) - F(P)}{X_t - P} \leq 1. \tag{13}$$

If we take the limit (as X_t approaches P) of the left hand side of this last inequality, we get $|F'(P)|$. If we require that $|F'(P)| < 1$, then the equality in (13) will be true for X_t close enough to P, so we have local stability at P.

Exercise 8. Use this result to show that the equilibrium point in our model is stable if $r < 2$.

Actually, the equilibrium point is *globally stable* for $r < 2$. That is, X_t approaches 1 as t approaches infinity even if it doesn't start close to 1. (For a proof of this, see [3].) Some pictures may help explain this behavior. In Fig. 3, $|F'(1)| < 1$ and X_t gradually moves toward 1. In Fig. 4, $|F'(1)| > 1$ and X_t gets further from 1.

Notice in Figs. 3 and 4 how we start with X_t on the x-axis, move vertically to find $F(X_t) = X_{t+1}$ on the graph, and then move horizontally to the line $y = x$ to find the corresponding point on the x-axis.

If we were to continue the process started in Fig. 4, we would find that X_t does not continue to get further from 1. To see what is going on, let us look at what happens over two generations by examining $F(F(X_t)) = X_{t+2}$. If we plot X_{t+2} vs. X_t for values of r slightly larger than 2, we will see something like Fig. 5, which uses $r = 2.3$. Let us call this new relation $F^{(2)}(X)$.

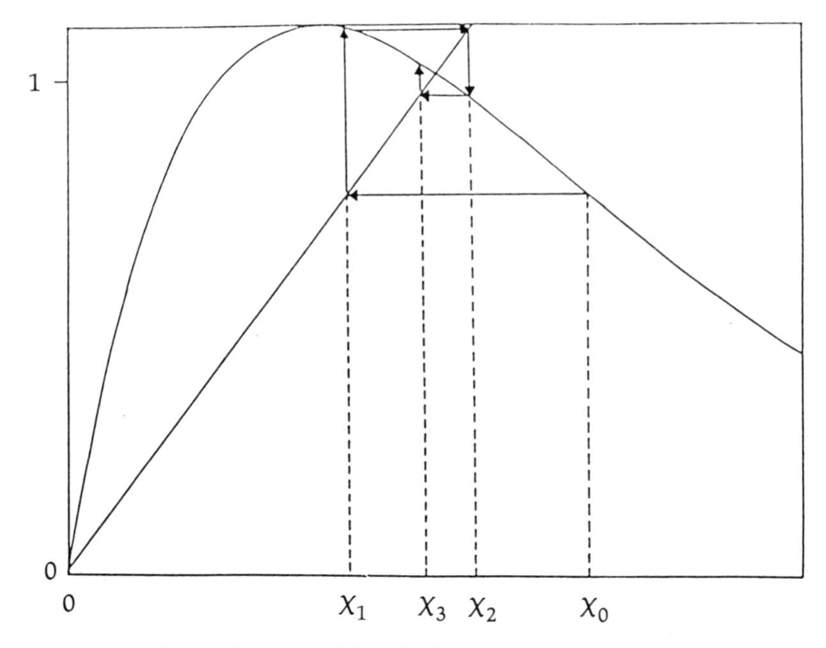

Figure 3. The population graph for $|F'(1)| < 1$.

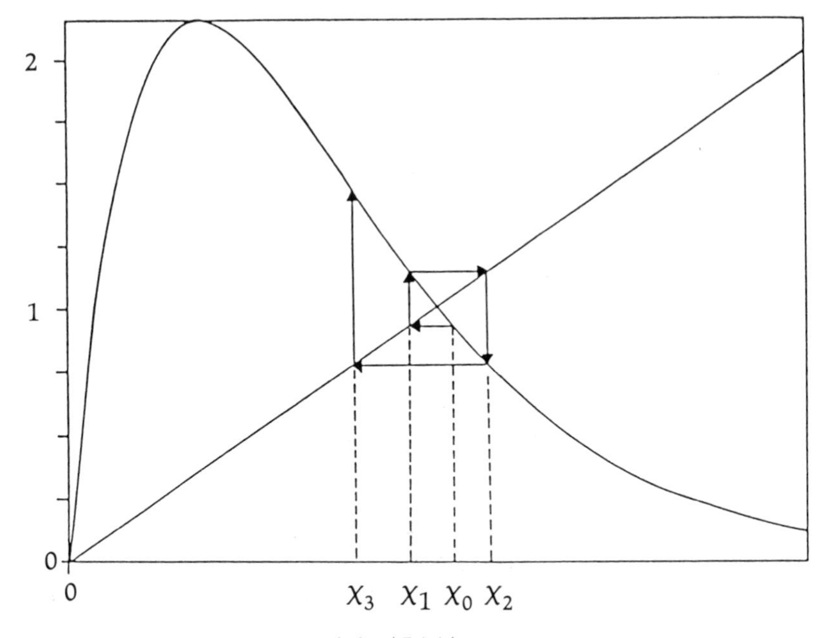

Figure 4. The population graph for $|F'(1)| > 1$.

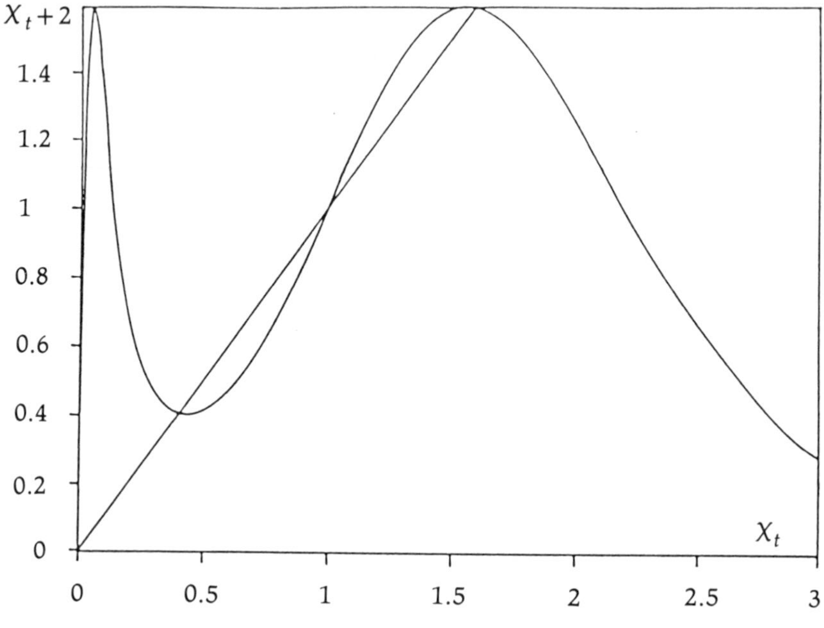

Figure 5. The graph of X_{t+2} vs. X_t.

Exercises **9.** Show that

$$F^{(2)}(x) = x\, e^{r(2-x-xe^{r(1-x)})}.$$

10. Show that an equilibrium of period 2 occurs, that is, a point where $F^{(2)}(x) = x$, if

$$2 - x - xe^{r(1-x)} = 0. \tag{14}$$

11. Find one solution to Eq. (14) by inspection. (If you think about what we have shown so far, you should be able to guess the answer.)

The other two solutions of Eq. (14) may be found by numerical methods. Using Newton's Method for example, you will find that with $r = 2.3$, the other two solutions are $x_1 = .4078$ and $x_2 = 1.592$. These two solutions, known as *periodic points of period 2*, are important because they truly have period 2, whereas $x = 1$ has period 1.

Using the result of Exercise 3, we find that $F'(x_1) = .242$ and $F'(x_2) = -.682$ (using x in place of N and letting $P = 1$). Since both of these are smaller in magnitude than 1, both points are stable. Together they form what is called a *limit cycle*. But as r gets bigger, x_1 and x_2 eventually succumb to the same fate as the original periodic point of period 1. When r becomes greater than 2.52, x_1 and x_2 become unstable and give rise to four stable periodic points of period 4.

The process, known as *bifurcation*, continues as r gets larger. The four periodic points of period 4 become unstable and bifurcate into eight periodic points of period 8, which become unstable and bifurcate into sixteen periodic points of period 16, and so forth. As r approaches a limiting value r_c, the bifurcations come faster and faster until finally, when $r > r_c$, the process becomes chaotic. In this region, there are an infinite number of periodic points, but these are usually unstable. It is important to realize that an unstable period will never be observed, since, if the population deviates even slightly from one of the periodic points, it will gradually drift farther away.

7. Computer Simulation

We can simulate the pattern of behavior described in the last section on a computer. The following BASIC program shows the change in population for any r over an arbitrary number of years:

```
10 INPUT "ENTER R, X, AND N: " ; R, X, N
20 FOR I = 1 TO N
30 PRINT I,X
40 X = X * EXP (R* (1-X))
50 NEXT I
60 END
```

When you run this program, the computer will ask you to input values of R (the parameter of r), X (the initial population), and N (the number of years you want to observe).

If you are using a computer with a monitor rather than a printer, you will want to know how to stop and restart the computations. Otherwise, if N is large, you will see a blur of numbers disappear off the top of the screen. (On an Apple microcomputer, the printing may be stopped and restarted by holding the CONTROL key down and pressing S.)

Exercises **12.** Run the program with a value of R between 2.0 and 2.5. Let the initial population be any number between 0 and 3. You should see the population approach a limit cycle of period 2 within 20 years.

13. Try the program with the same R as in Exercise 12 but a different initial population. What happens? Can you explain this?

14. Run the program with R = 2.55. You should see a limit cycle with period 4.

15. Run the program with R = 2.7. Do you see any limit cycle? If you think you have found one, try making N really large (try 200) and see whether the pattern continues.

As you should have found in Exercise 14 and 15, r_c is somewhere between 2.5 and 2.7. The correct value, up to four decimal places, is 2.6924. (To see how this may be calculated, see the article by May and Oster [8].)

Exercise **16.** Run the program with R = 3.12 and N = 40. You should observe a limit cycle with period 3.

The cycle of period 3 you observed in Exercise 16 arises when r is large enough for there to be solutions of the equation $F(F(F(x)))$ $= x$ other than $x = 1$.

Exercise **17.** Show that this last equation holds if

$$3 - x[1 + e^{r(1-x)} + e^{r(2-x(1+e^{r(1-x)}))}] = 0. \tag{15}$$

In Fig. 6, we see the graph of the function found in Exercise 17 for $r = 3.1024$. This is the smallest value of r for which Eq. (15) has a solution other than $x = 1$. Once r becomes large enough, this cycle of period 3 also becomes unstable.

Exercise **18.** Verify this last statement by running the program with R = 3.5.

Tien-Yien Li and James A. Yorke [7] have shown that if a model in Eq. (9) has a cycle of period 3, then it has cycles of period k for any positive integer k. Even if these cycles were stable, we would not be able to observe those with large periods unless we watched the population for a great number of years. In other words, a cycle that repeats every 1000 years is indistinguishable from a random sequence of numbers if we observe the population for only 50 years.

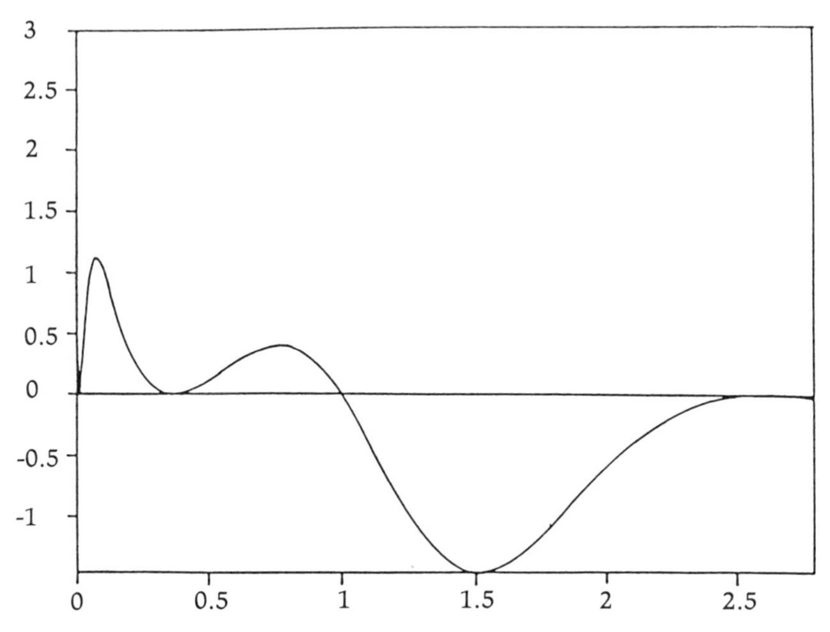

Figure 6. The graph of the left-hand side of Eq. (15) as a function of x for $r = 3.1024$.

The amazing thing is that such a simple model as Eq. (9) can exhibit such bizarre behavior. If you observed the data from Exercise 18 without knowing where it came from, you would be unlikely to guess the correct model. Perhaps even more amazing is that this behavior occurs for *any* model that has a hump-shaped graph as in Figure 1. You may wish to see the articles by Li and Yorke and by May and Oster for a more detailed explanation.

8. Dynamic Programming

For a variety of reasons, the strategy of catching the maximum sustainable yield is not necessarily best. Putting ourselves in the fishermen's waders, why do we want to keep the catch constant for all eternity if we won't be around long enough to fish it? Furthermore, might we be better off in the long run if we allow the catch to vary from one year to the next?

To pursue answers to those questions, imagine that we have a generation of salmon swimming around us, which we will designate generation 0. One action we can take is to catch none of them and let them spawn, hoping we will get a larger population next year. The opposite extreme is to catch them all right now. Or we can choose the intermediate action of catching a fraction of the population. Our goal is to maximize the total catch from generation 0 through generation n, when n is some positive integer (perhaps the number of years we expect to be in business).

One way to tackle this problem is to use the technique of dynamic programming. Despite the formidable name, the reasoning behind dynamic programming is fairly simple. Suppose we have generation i in front of us, where $0 \leq i \leq n$ and we decide to catch the fraction u_i of X_i, where $0 \leq u_i \leq 1$. Such a u_i is called the *exploitation rate*. After the catch, we only have

$$S_i = X_i(1-u_i), \tag{17}$$

the so-called spawning population, left to give birth to the next generation. Consequently, instead of Eq. (9), we now have

$$X_{i+1} = S_i e^{r(1-S_i)}. \tag{18}$$

For any u_i chosen between 0 and 1, the catch is u_iX_i (relative to the equilibrium population), and the next generation is X_{i+1}, as given by Eq. (18).

If we let $g(X_i)$ denote the maximum total catch from generations i through N, given a generation i population of X_i, we will get a maximum of $u_iX_i + g(X_{i+1})$ from generation i through N if we choose the exploitation rate u_i right now. We therefore choose the u_i that maximizes the quantity $u_iX_i + g(X_{i+1})$. In other words,

$$g(X_i) = \max_{0 \leq u_i \leq 1} \{u_iX_i + g(X_{i+1})\}, \tag{19}$$

with X_{i+1} given by Eqs. (17) and (18).

The trick of dynamic programming is to use Eqs. (17), (18), and (19) to work backwards from the last generation to the current generation. If we are already at generation n, what is the best strategy? Making the assumption that we cannot profit from any fish left behind after this year, then we want to catch all the fish available, so $g(X_n) = X_n$. (Good for us, bad for any future generations of salmon lovers or fishermen.) For generation $n-1$, using Eqs. (17), (18), and (19),

$$
\begin{aligned}
g(X_{n-1}) &= \max_{0 \le u_{n-1} \le 1} \quad u_{n-1} X_{n-1} + g(X_n) \\
&= \max_{0 \le u_{n-1} \le 1} \quad u_{n-1} X_{n-1} + X_n \\
&= \max_{0 \le u_{n-1} \le 1} \quad u_{n-1} X_{n-1} + S_{n-1} e^{r(1-S_{n-1})} \\
&= \max_{0 \le u_{n-1} \le 1} \quad u_{n-1} X_{n-1} + X_{n-1}(1-u_{n-1})\, e^{r(1-X_{n-1}(1-u_{n-1}))}
\end{aligned}
\tag{20}
$$

The usual way of getting u_{n-1} from Eq. (20) is the discretization technique. Instead of regarding u and X as continuous variables, we regard them as discrete variables, taking on only a finite number of discrete values. For each possible X_{n-1}, we choose the u_{n-1} which maximizes Eq. (20), using a computer to perform the calculations. Then we consider all the possible values of X_{n-2}. For each value, we choose u_{n-2} to maximize Eq. (19) with $i = n-1$.

We continue working our way backwards in this manner until we reach X_0. We know what X_0 is, and from the previous work we know what u_0 should be for this value of X_0. Then Eqs. (17) and (18) tell us X_1, and from the previous work we find the corresponding value of u_1. We continue to work forward until we reach generation n. To go into the details is beyond the scope of this unit, but interested readers may wish to investigate [14] and [15]. Let's just say that dynamic programming is an elegant and efficient (ef-fish-ient?) way to solve the problem.

Exercise 19. Suppose fishing regulations do not allow us to catch all the salmon in generation n, but instead require us to leave at least L salmon behind, or, if X_n is less than L, to leave behind all of X_n. Write a corresponding expression for $g(X_n)$.

Often next year's catch is more important to the fishing industry than the catch, say, ten years from now. For one thing, the distant future is too unpredictable. Another reason is that, if we catch more fish in the short term, the money we make from fishing could be invested elsewhere (perhaps in a Swiss bank account) to further increase our earnings. Thus, some biologists introduce a discount factor, v, into the catches, where $0 < v < 1$. The assumption is usually made that each year's catch is valued at a constant fraction v of the previous year's catch. Then, instead of trying to maximize the total catch

$$\sum_{i=0}^{n} u_i X_i,$$

we try to maximize the total discounted catch

$$\sum_{i=0}^{n} v^i u_i X_i.$$

Exercises **20.** If $0 < v < 1$, prove that the sum

$$\sum_{i=0}^{n} v^i u_i X_i$$

puts more weight on the near future than the distant future.

21. Modify the dynamic programming formulation, Eq. (19), to maximize the total discounted catch.

9. Other Models

The Ricker model is not the only model used to study salmon. For example, another one developed by Beverton and Holt [2] yields the equation

$$N_{t+1} = \frac{1}{a + b/N_t} \quad . \tag{21}$$

The articles by May and Oster [8] and by Lamberson and Biles [6] list related models.

10. References

In addition to the books and articles referred to in the text, we have included a few other references [4, 5, 9, 10] that may be of interest.

1. Alexander, John W., *Curve Fitting via the Criterion of Least Squares*, UMAP Unit 321.

2. Beverton, R. J. H. and Holt, S. J., "On the dynamics of exploited fish populations," *Fishery Investigation of the Ministry of Agriculture Fisheries and Food* (Great Britain), Series II, Vol. 19 (1957), pp. 1-533.

3. Fisher, M. E., Goh, B. S. and Vincent, T. L., "Some stability conditions for discrete-time single species models." *Bulletin of Mathematical Biology*, Vol. 41 (1979), pp. 861-875.

4. Greenwell, Raymond N., "Whales and krill: a Mathematical model," *The UMAP Journal*, Vol. 3 (1982), pp. 165-183. Also published as UMAP Unit 610. This module develops a mathematical model for another ecological process and makes additional references to books and articles on mathematical ecology.

5. Jones, J. W., *The Salmon*, Harper and Brothers, 1959. This book, [9], and [10] provide a great deal of information on the life and times of salmon.

6. Lamberson, Roland, and Biles, Charles, "Polynomial models of biological growth," *The UMAP Journal*, Vol. 2, No. 2 (1981), pp. 9-25.

7. Li, Tien-Yien and Yorke, James A., "Period Three Implies Chaos," *American Mathematical Monthly*, Vol. 82 (1975), pp. 985-992.

8. May, Robert M. and Oster, George F., "Bifurcations and dynamic complexity in simple ecological models," *The American Naturalist*, Vol. 110 (1976), pp. 573-599.

9. Mills, Derek, *Salmon and Trout: a Resourcę, its Ecology, Conservation and Management*, St. Martin's Press, 1971.

10. Netboy, Anthony, *The Salmon: Their Fight for Survival*, Houghton Mifflin Company, 1974.

11. Ricker, W. E., "Stock and Recruitment," *Journal of the Fisheries Research Board of Canada*, Vol. 11 (1957), pp. 559-623.

12. Shepard, M. P. and Withler, F. C., "Spawning stock size and resultant production for Skeena sockeye," *Journal of the Fisheries Research Board of Canada*, Vol. 15 (1958), pp. 1007-1025.

13. Thomas, William P., Pomerantz, Mark J. and Gilpin, Michael E., "Chaos, asymmetric growth and group selection for dynamic stability," *Ecology*, Vol. 61 (1980), pp. 1312-1320.

14. Walters, Carl J., "Optimal harvest strategies for salmon in relation to environmental variability and uncertain production parameters," *Journal of the Fisheries Research Board of Canada*, Vol. 32 (1975), pp. 1777-1784.

15. Walters, Carl J., "Optimum escapements in the face of alternative recruitment hypotheses," *Canadian Journal of Fisheries and Aquatic Sciences*, Vol. 36 (1981), pp. 678-689.

11. Answers to Exercises

1. $N_{t+1} = kN_t$, so $N_1 = kN_0$, $N_2 = kN_1 = k^2N_0$, $N_3 = kN_2 = k^3N_0$, and, in general, $N_t = N_0k^t$. Mathematical induction could be used to make this more rigorous.

2. $$\frac{dR}{dt} = -CR.$$

   Separating variables and integrating both sides, we have

   $$\frac{dR}{dt} = \quad -C\,dt,\ ln\,R = -Ct + K,\ R = e^{-Ct}\,e^K.$$

   When $t = 0$, $R = e^K$, so we denote e^K by R_0. Thus $R = R_0\,e^{-Ct}$.

3. $f'(N) = e^{r(1-N/P)} + N e^{r(1-N/P)}(-r/P) = (1-r N/P) e^{r(1-N/P)}$.

4. $$\lim_{N\to\infty} N e^{r(1-N/P)} = \lim_{N\to\infty} \frac{e^r N}{e^{r N/P}}.$$

   Since the numerator and denominator approach ∞, we invoke L'Hopital's rule, yielding

   $$\lim_{N\to\infty} \frac{e^r}{e^{r N/P} \cdot (r/P)} = 0$$

5. To maximize $F(N) - N$, set the derivative equal to 0: $f'(N) - 1 = 0$ or $f'(N) = 1$.

6. Since $F'(N^*) = 1$, the answer to Exercise 3 tells us that

   $$(1-rN^*/P)\, e^{r(1-N^*/P)} = 1,$$

   $$e^{r(1-N^*/P)} = 1/(1-rN^*/P).$$

   The maximum sustainable yield is

   $$f(N^*) - N^* = N^* e^{r(1-N^*/P)} - N^* = N^*\left(\frac{1}{1-rN^*/P} - 1\right)$$

7. $$N_{t+1}/N_t = e^{r(1-N_t/P)},$$
   $$\text{so } y_t = ln(N_{t+1}/N_t)$$
   $$= r(1-N_t/P)$$
   $$= r - N_t r/P,$$

   which is the equation of a line with slope $-r/P$ and y-intercept r.

8. $F'(X) = (1-rN)\, e^{r(1-N)}$. Since $P = 1$,

   $$|F'(P)| < 1,\ |1-r| < 1,\ 0 < r < 2.$$

   We have already assumed $r > 0$, so we only need $r < 2$.

9. $$F^{(2)}(X) = F(F(X))$$
   $$= F(X)\, e^{r(1-F(X))}$$
   $$= xe^{r(1-x)} e^{r(1-Xe^{r(1-x)})}$$
   $$= xe^{r(1-x+1-xe^{r(1-x)})}$$
   $$= xe^{r(2-x-xe^{r(1-x)})}.$$

10. If $x = F^{(2)}(x) = xe^{r(2-x-xe^{r(1-x)})}$, then $e^{r(2-x-xe^{r(1-x)})} = 1$, so $2-x-xe^{r(1-x)} = 0$.

11. $x = 1$ is a solution.

12. The limit cycle depends on the value of r chosen. If $r = 2.2$, a limit cycle consisting of the two points .49706 and 1.50294 is seen by the time $N = 20$.

13. Since the periodic points are globally stable, we should see the same limit cycle regardless of the starting point.

14. The cycle consists of the four points .22037, 1.60900, .34050, 1.83012.

15. There is no stable limit cycle. Any one that seems to appear will drift if N gets large enough.

16. The limit cycle consists of the three points .01458, .3155, and 2.670.

17. Using the result of Exercise 10,

$$\begin{aligned} F^{(3)}(x) &= F(F^{(2)}(x)) = F^{(2)}(x)\, e^{r(1-F^{(2)}(x))} \\ &= xe^{r(2-x-xe^{r(1-x)})} e^{r(1-xe^{r(2x-xe^{r(1-x)})})} \\ &= xe^{r(2-x-xe^{r(1-x)}+1-xe^{r(2-x-xe^{r(1-x)})})} \\ &= xe^{r(3-x-xe^{r(1-x)}-xe^{r(2-x-xe^{r(1-x)})})}. \end{aligned}$$

This is equal to x if $3-x-xe^{r(1-x)}-xe^{r(2-x-xe^{r(1-x)})} = 0$, or $3-x(1+e^{r(1-x)}+e^{r(2-x-xe^{r(1-x)})}) = 0$.

18. There is no stable limit cycle. The points appear to be random.

19. $g(X_n) = \max\{X_n - L, 0\}$.

20. If $m < n$, then $v^m > v^n$ since $0 < v < 1$.

21. $g(X_i) = \max\limits_{0 \le u_i \le 1} \{u_i X_i + vg(X_i + 1)\}$.

UMAP

Modules in Undergraduate Mathematics and its Applications

Module 657

Controlling the Effects of Interruption

Jo Anne Growney

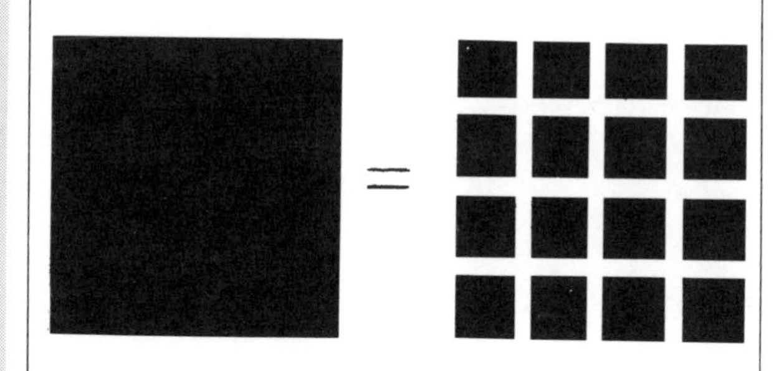

Published in cooperation with the Society for Industrial and Applied Mathematics, the Mathematical Association of America, the National Council of Teachers of Mathematics, the American Mathematical Association of Two-Year Colleges, and The Institute of Management Sciences.

INTERMODULAR DESCRIPTION SHEET:	UNIT 657
TITLE:	Controlling the Effects of Interruptions
AUTHOR:	Jo Anne Growney Department of Mathematics and Computer Science Bloomsburg University Bloomsburg, PA 17815
MATH FIELD:	Discrete Mathematics
APPLICATION FIELD:	Scheduling
TARGET AUDIENCE:	Students in finite mathematics, mathematical modeling or management science courses.
ABSTRACT:	This module deals with the problem of scheduling a task when interruptions occur with a known probability. Numerous examples demonstrate the benefits of scheduling and of organizing tasks into manageable subtasks.
PREREQUISITES:	Elementary notions of probability.

The UMAP Journal, Vol. V, No. 3, 1984

Controlling the Effects of Interruptions

Jo Anne Growney
Department of Mathematics and Computer Science
Bloomsburg University
Bloomsburg, PA 17815

Table of Contents

1. INTRODUCTION 1
2. SIMULATION OF INTERRUPTIONS 2
3. MORE INTERRUPTION PROBLEMS—AND A FORMULA 5
4. THE VALUE OF ORGANIZATION 10
5. DERIVATION OF INTERRUPTION-TIME FORMULA 19
6. SCHEDULING AROUND INTERRUPTIONS 21
7. MORE EXERCISES 25
8. HINTS AND ANSWERS FOR EXERCISES 26

Modules and Monographs in Undergraduate Mathematics and its Applications Project (UMAP)

The goal of UMAP was to develop, through a community of users and developers, a system of instructional modules in undergraduate mathematics and its applications to be used to supplement existing courses and from which complete courses may eventually be built.

The Project was guided by a National Advisory Board of mathematicians, scientists, and educators. UMAP was funded by a grant from the National Science Foundation and is now supported by the Consortium for Mathematics and Its Applications, Inc. (COMAP), a nonprofit corporation engaged in research and development in mathematics education.

The project would like to thank Paul Nugent of Franklin College and Harvey Braverman of New York City Technical College for their reviews, and all those who assisted in the production of this unit.

This material was prepared with the partial support of National Science Foundation Grant No. SPE8304192. Recommendations expressed are those of the authors and do not necessarily reflect the views of the NSF or the copyright holder.

Some of the ideas for this module appeared in Vol. #55 #4 (Sept. 1982) *Mathematics Magazine*, "Planning for Interruptions", pp. 213-219.

1. Introduction

Gwynne is a residence counselor in a freshman dorm. Her study time is interrupted frequently by knocks on her door and questions from students who live around her. Yet Gwynne always has her class assignments completed on time. When asked how she manages to get her work done despite the many interruptions, Gwynne responded with a simple rule-of-thumb:

"After I estimate how much time I think an assignment will take, I double my estimate. I then schedule the double time to get everything done."

Gwynne is not alone in her need to schedule more time for a task than she really thinks it will take in order to allow for interruptions. Lots of people—students, teachers, managers, parents—spend much of their time in situations in which tasks cannot be completed without interruptions. Some even find themselves with projects that never get completed because interruptions break concentration and cause such setbacks that continual starting over is required. Often, as in Gwynne's case, dealing with interruptions is part of a person's responsibility and locking the door on interruptions is not a reasonable alternative.

But it also does not seem reasonable that in order to get jobs done we should need to allow, as Gwynne does, twice as much time as the job really needs.

In this module we shall consider another way to deal with setbacks caused by interruptions, namely organization. If a person is willing to invest time in organizing tasks, then he or she can reduce the amount of time wasted by interruptions.

We begin by examining a simple example, one of a task beset with interruptions, that can be illustrated easily by simulation.

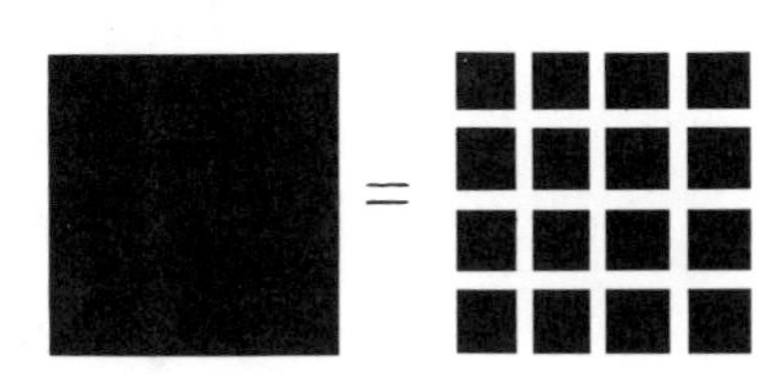

2. Simulation of Interruptions

The following example is definitely make-believe but provides a model for later realistic examples.

Example 1.

Paul has become aware of a ring of flab around his waist and of other signs of physical unfitness. He has resolved to return to being trim and fit. Being a sensible fellow, Paul consults his physician before embarking on a fitness program. The doctor recommends a daily stint of five uninterrupted minutes on a treadmill. Should Paul be interrupted during his five-minute effort he must start over and complete five additional minutes without interruption.

Sounds simple! However Paul has another difficulty. He has a chronic skin disorder. Randomly, but on the average of once every five minutes, Paul is interrupted by the necessity to stop to scratch an itch. No matter how hard he scratches at any given time, Paul cannot prevent the random and frequent reoccurrence of this annoyance.

How does Paul fare with his fitness program?

Simulation:

If Paul's itches occur randomly but on the average of once every five minutes then we may describe this by saying that he has probability

$$p = \frac{1}{5}$$

of interruption during any given minute.

We can simulate Paul's experience on the treadmill using the table of random digits in Fig. 1.

Starting at any point in the table we can read digits and interpret each digit as a minute on the treadmill. If the digits 1, 2, 3, 4, 5, 6, 7 and 8 denote minutes in which no itch interrupts and if the digits 9 and 0 denote interrupted minutes, then the relative frequency of interruptions will be 2/10 or 1/5. A five-minute uninterrupted jog will be simulated when we encounter five digits from 1, 2, 3, 4, 5, 6, 7, 8 in a row.

Row 1	86298	26610	90511	08055	80364	70233	91451	34528	30357	27456
	93680	27051	67692	57437	08779	81065	50586	20621	28296	43353
	45153	17985	74725	08526	09220	89778	59841	02387	78112	16035
	65055	40547	20834	50243	23998	59708	12313	89349	25103	43682
	80863	76681	73173	48970	91202	81344	89446	60285	12653	95567
Row 6	65704	35329	80233	67505	22518	58994	63968	79316	53447	65610
	16862	82356	69963	61171	96043	56593	73637	82198	51634	71363
	76048	34462	57543	98743	80838	42517	42094	98970	07496	22223
	92003	32221	39595	99113	43596	90842	87684	80098	54888	32782
	74244	90661	80795	20305	92055	54532	99534	34660	41569	88305
Row 11	38128	35924	55245	97971	52694	92422	15875	18971	20058	78333
	33729	56998	99535	52712	21588	36734	24131	95807	80922	85010
	63971	68875	13322	07349	73991	41072	31419	29611	10297	85465
	57653	56330	22804	71402	62635	33217	85828	69039	77095	57603
	36395	30423	96224	53481	23420	44921	30883	56083	32038	63699
Row 16	90543	52660	09346	76795	89783	87944	92379	34576	18055	67418
	58133	19098	70130	16092	43843	80508	96387	42270	35335	18264
	57487	88972	50914	65331	87902	42601	85407	19867	77391	48159
	77128	23219	48346	02047	63984	66444	83317	40167	39020	00798
	13964	87042	24341	25448	30779	30472	92064	71532	47311	33061

Figure 1. A table of random digits is a listing of the digits 0, 1, 2, 3, 4, 5, 6, 7, 8 and 9 with the property that in any position of the table each of the digits is equally likely to appear. The table shown here contains 1000 such random digits.

If we start the simulation with the first digit of the first row of Fig. 1, then we have the following results:

Digit	**Interpretation**
8	1st minute of jogging completed
6	2nd minute of jogging completed
2	3rd minute of jogging completed
9	Stop to itch; start-over required
8	1st minute of jogging completed
2	2nd minute of jogging completed
6	3rd minute of jogging completed
6	4th minute of jogging completed
1	5th minute of jogging completed

In this case we see that the effort to complete five uninterrupted minutes of jogging required nine minutes altogether.

Of course, one simulation experiment provides only a sample of what can happen. To determine what can be expected on the average we need to perform many simulations and find the mean of the lengths of time that the individual simulations require.

Suppose we decide to perform 25 repetitions of the experiment that simulates Paul's jogging effort. Figure 2 indicates the blocks of random digits that constitute the simulations. Each group of five digits marked with ______| designates the five uninterrupted minutes that complete Paul's jogging effort.

Row 1	86298	26610	90511	08055	80364	70233	91451	34528	30357	27456
	93680	27051	67692	57437	08779	81065	50586	20621	28296	43353
	45153	17985	74725	08526	09220	89778	59841	02387	78112	16035
	65055	40547	20834	50243	23998	59708	12313	89349	25103	43682
	80863	76681	73173	48970	91202	81344	89446	60285	12653	95567
Row 6	65704	35329	80233	67505	22518	58994	63968	79316	53447	65610
	16862	82356	69963	61171	96043	56593	73637	82198	51634	71363
	76048	34462	57543	98743	80838	42517	42094	98970	07496	22223
	92003	32221	39595	99113	43596	90842	87684	80098	54888	32782
	74244	90661	80795	20305	92055	54532	99534	34660	41569	88305
Row 11	38128	35924	55245	97971	52694	92422	15875	18971	20058	78333
	33729	56998	99535	52712	21588	36734	24131	95807	80922	85010
	63971	68875	13322	07349	73991	41072	31419	29611	10297	85465
	57653	56330	22804	71402	62635	33217	85828	69039	77095	57603
	36395	30423	96224	53481	23420	44921	30883	56083	32038	63699
Row 16	90543	52660	09346	76795	89783	87944	92379	34576	18055	67418
	58133	19098	70130	16092	43843	80508	96387	42270	35335	18264
	57487	88972	50914	65331	87902	42601	85407	19867	77391	48159
	77128	23219	48346	02047	63984	66444	83317	40167	39020	00798
	13964	87042	24341	25448	30779	30472	92064	71532	47311	33061

Figure 2. Each block underlined here describes five uninterrupted minutes in the simulation. Each new experiment begins with the table digit that follows the last digit of the previous experiment.

Figure 3 tabulates the number of minutes needed to complete each experiment. Although the lengths of time required vary from five to 28 minutes, the mean length of time for the 25 experiments is approximately 11.7 minutes.

Simulation Experiment #	Total number of minutes needed to include five uninterrupted minutes
1	9
2	27
3	5
4	6
5	16
6	6
7	23
8	7
9	5
10	9
11	28
12	5
13	26
14	12
15	15
16	8
17	5
18	5
19	12
20	13
21	9
22	8
23	8
24	7
25	18
Mean length for 25 simulation experiments is 292/25 ≃ 11.7 minutes.	

Figure 3. Lengths of 25 experiments simulating Paul's jogging effort.

Exercise 1.

Continuing from the point in Fig. 2 where we finished the 25th simulation experiment, perform 25 more simulations of Paul's jogging effort. Record your results in a continuation of Fig. 3. Find the mean length of the 50 experiments.

3. More Interruption Problems— and a Formula

The example of Paul's jogging was unrealistic and somewhat silly. But it can serve as a springboard to discussion of problems that are more realistic and more important.

We shall consider problems with the following characteristics:

> A person has a task that requires a number U of uninterrupted time units to complete.
>
> The person is subject to interruptions during the task; the probability of interruption during any given time unit is p.
>
> Interruptions are independent. That is, the occurrence of one interruption has no effect on the probability of subsequent interruptions.
>
> Interruptions are devastating. That is, every time an interruption occurs the person must start the task over.

As students, we encounter many problems that fit these characteristics at least roughly.

> Tasks like writing papers and solving complex math problems require periods of uninterrupted concentration.
>
> Often we must perform tasks at times and in places in which there is a likelihood of random interruption; music, visitors, phone calls, even fatigue and hunger can destroy our concentration. If we can estimate an average number i of time units between interruptions then the probability p of interruption during any given time unit is $p = 1/i$.
>
> Independence of interruptions is hard to determine. If our interruptions are completely outside our control it may be reasonable to suppose that they are independent.
>
> Sometimes, when our concentration on a task is interrupted, our progress is not wholly devastated and we need not start at the beginning after the interruption. Often, though, interruptions are accompanied by a build-up of frustration and a fatigue effect that are almost equivalent to starting over.

In situations of the sort just described, in which a task requires U uninterrupted time units , and interruptions that are independent and devastating occur randomly with probability p, then the expected (or average) length of time T required to complete the task is given by the following formula:

$$T = \frac{1}{p}\left[\frac{1}{(1-p)^U} - 1\right] . \qquad (1)$$

A derivation of Eq. (1) is outlined in Section 5 of this module.

In Example 1, which describes Paul's physical fitness activity, $U = 5$ minutes and $p = 1/5 = 0.2$. Substitution of these values into Eq. (1) leads to

$$T = \frac{1}{p}\left[\frac{1}{(1-p)^U} - 1\right].$$

$$= \frac{1}{0.2}\left[\frac{1}{(.8)^5} - 1\right]$$

$$= 5\left[\frac{1}{0.32768} - 1\right]$$

$$\simeq 5[3.052 - 1]$$

$$\simeq 10.3 \text{ minutes.}$$

Although we have not yet derived Eq. (1), the simulation completed in Exercise 1 provides a check on the reasonableness of the equation. The average for these fifty simulation experiments is just under 10.5 minutes, a result that is close to the answer obtained with the equation.

Several exercises below provide practice with calculations using Eq. (1). We shall then go on to show how the equation can be used in evaluating different scheduling options.

Exercise 2.

Read all three problems *a*, *b*, and *c* before solving any of them. Note their similarity and, before using Eq. (1) to calculate values for T, guess an answer to each.

a) Max has a job that requires ten uninterrupted minutes. When an interruption occurs, he must start over. Max's interruptions occur randomly and independently and, on the average, once every five minutes. Thus the probability of an interrupton during any given minute is $p = 1/5 = 0.2$. What is the expected length of time that Max's job will require?

b) Marsha has a job that requires ten uninterrupted minutes. When an interruption occurs, she must start over. Marsha's interruptions occur randomly and independently and, on the average, once every ten minutes. What is the expected length of time that Marsha's job will require?

c) Melanie has a job that requires ten uninterrupted minutes. When an interruption occurs, she must start over. Melanie's interruptions occur randomly and independently and, on the average, once every 20 minutes. What is the expected length of time that Melanie's job will require?

d) Compare the answers to (**a**), (**b**) and (**c**) with each other and with your guesses. Are the results surprising? Why is Max's expected time so long?

Exercise 3.

Mike and Michelle have a talent for building houses of cards. A house of cards is a structure built from ordinary playing cards, stood on edge, using only each other for support. This coming weekend they want to break their old record, which is a ten-story house. But they have an interruption problem in the form of their younger brother Mitch who keeps coming around their work table, either breaking someone's concentration or creating a breeze or vibration that brings the house down. Mike and Michelle estimate that they will need at least one uninterrupted hour (60 minutes) to build a house of cards that is higher than ten stories.

a) If Mitch can be induced to watch TV, Mike and Michelle estimate that he will cause interruptions randomly and independently and, on the average, about once every 15 minutes. With this interruption pattern, what is the expected length of time that it will take for Mike and Michelle to build a record-breaking house of cards?

b) If the weather is nice on Saturday, Mitch will spend most of his time outside. In this case he will interrupt only, on average, once every hour. With this interruption pattern, what is the expected length of time that it will take for Mike and Michelle to build their record-breaking house of cards?

c) Based on the answers you obtained in (**a**) and (**b**), what advice would you give to Mike and Michelle?

Exercise 4.

The formula for T (the total time that a task requires because of startovers due to interruptions) depends on two quantities:

U – the number of uninterrupted time units that the task requires, and

p – the probability of interruption during any given time unit.

To become familiar with exactly how changes in U and p affect T, it is helpful to perform a series of related calculations. To guide you in doing this, selected values of U and p are supplied in the tables (**a**), (**b**), and (**c**) below. Use Eq. (1) to calculate the value of T in each case. Items (**d**)-(**i**) ask questions that will help you to analyze table values.

(**a**)

U	5 min.	5 min.	5 min.	5 min.	5 min.	5 min.	5 min.
p	$\frac{1}{100} = 0.01$	$\frac{1}{20} = 0.05$	$\frac{1}{10} = 0.1$	$\frac{1}{5} = 0.2$	$\frac{1}{4} = 0.25$	$\frac{1}{3} \simeq 0.33$	$\frac{1}{2} = 0.5$
T							

(**b**)

U	10 min.	10 min.	10 min.	10 min.	10 min.	10 min.	10 min.
p	$\frac{1}{100} = 0.01$	$\frac{1}{20} = 0.05$	$\frac{1}{10} = 0.1$	$\frac{1}{5} = 0.2$	$\frac{1}{4} = 0.25$	$\frac{1}{3} \simeq 0.33$	$\frac{1}{2} = 0.5$
T							

(**c**)

U	20 min.	20 min.	20 min.	20 min.	20 min.	20 min.	20 min.
p	$\frac{1}{100} = 0.01$	$\frac{1}{20} = 0.05$	$\frac{1}{10} = 0.1$	$\frac{1}{5} = 0.2$	$\frac{1}{4} = 0.25$	$\frac{1}{3} \simeq 0.33$	$\frac{1}{2} = 0.5$
T							

Use the tables (**a**), (**b**), *and* (**c**) *to answer the following questions.*

d) For a fixed value of U, how are the values of T and p related?

e) Suppose George has an "interruption tolerance" of 50%; that is, he is unwilling to have T exceed U by more than 50% of U.

Should George attempt to complete a five-minute uninterruptible task if the probability of being interrupted during any given minute is $p = 0.25$? What if $p = 0.2$? f $p = 0.1$? If $p = 0.05$? Should he attempt to complete a ten-minute uninterruptible task when $p = 0.2$? When $p = 0.1$? When $p = 0.05$?

f) The formula $i = 1/p$ gives the expected length i of the interval of time between interruptions. Observe that $T \cong 2U$ in the table of (**a**) when $U = i$. Can the same observation be made in the tables of (**b**) and (**c**)?

Might most people intuitively expect that when $U = i$ the value of T will be close to U? Supply a verbal (rather than mathematical) explanation to convince such individuals that their intuitive expectation is incorrect.

g) If the probability of interruption in any given minute is $p = 0.25$, use values from the tables of (**a**) and (**b**) to answer the following question. Which will take longer: three five-minute uninterruptible tasks or one ten-minute uninterruptible task?

h) If the probability of interruption in any given minute is $p = 0.1$, which will take longer: three ten-minute uninterruptible tasks or one twenty-minute uninterruptible task?

i) Provide an explanation for the "surprising" answers to (**g**) and (**h**). Invent other questions similar to these two and determine the answers. Assess your calculations and comparisons to develop a feel for the relationships between p, U and T.

4. The Value of Organization

Sometimes we are reluctant to spend time to organize the details of a task. For example, when an English composition is due we want simply to write it and be done without first taking time to organize our thoughts with an outline and then taking more time for the writing.

Waiting to start a task until we have first taken time to organize it is an approach that many of us resist. We question whether that extra time to get organized can possibly save time overall. Perhaps calculations you have already done convince you of the usefulness of organization as a time-saving strategy. Example 2 is designed to help convince you further; it describes a situation for which this approach is definitely beneficial.

Example 2.

Two political science students, Jim and Jean, have each been assigned to write a brief paper that analyzes and compares different voting methods. Each estimates that the rough draft will take one hour of intense concentration—that is, it must be done without interruptions. Any interruption will be so devastating that starting over will be required. For both Jim and Jean interruptions are independent and occur randomly—on the average of once every 20 minutes.

a) Jim started writing his rough draft at 6 p.m. last evening. Under the given conditions, what is the expected time at which he finished?

Analysis:
The uninterrupted time required is

$$U = 60 \text{ minutes.}$$

The likelihood of interruption during any minute is

$$p = \frac{1}{20} = 0.05.$$

Substitution of these values into Eq. (1) gives

$$T = \frac{1}{0.05}\left[\frac{1}{(1 - 0.05)^{60}} - 1\right] \simeq 414 \text{ minutes.}$$

Thus Jim's rough draft required almost seven hours of writing effort. If his interruptions were brief and did not add significantly to this time, his expected completion time was around 1 a.m. (It is important to note that the formula for T in Eq. (1) measures "time-on-task" and does not include time actually spent dealing with lengthy interruptions. As when we simulated Paul's jogging effort in Example 1, so also when we use Eq. (1) for T, we keep track of time spent working on the desired task and do not include time spent on interruptions.)

b) Jean was astounded when a sleepy-eyed Jim told her how long it had taken him to complete his rough draft. Because of his experience, Jean decided to try a different approach. Jean believes that with 15 minutes of intense concentration she can outline her paper and divide it into four sections, each of which will take 15 minutes to write. If she is interrupted during any 15-minute section, she will need to start over and spend 15 more minutes on that portion of her task. However, once her outline or any one of the four subsections is complete, no interruption will affect it.

Jean thus has five 15-minute tasks to complete. How long

will this take her? If she begins at 6 p.m. tonight, when can she expect to finish?

Analysis:
The uninterrupted time required is

$$U = 15 \text{ minutes.}$$

The likelihood of interruption during any given minute is

$$p = \frac{1}{20} = 0.05.$$

Substitution of these values into Eq. (1) gives

$$T = \frac{1}{0.05}\left[\frac{1}{(1 - 0.05)^{15}} - 1\right] \simeq 23.2 \text{ minutes.}$$

This value for T is the expected time required for Jean to complete one 15-minute task. Since she has set up five tasks for herself, the total time that Jean's rough draft will require is

$$5T \simeq 116 \text{ minutes.}$$

Jean's rough draft will thus require almost two hours for completion. If she starts at 6 p.m. this evening, she can expect to be done around 8 p.m. (or later if the actual time needed to deal with the interruptions is substantial).

Despite our calculations, it is hard to believe that Jim would spend seven hours on a one-hour paper. We can picture him, after a while, placing a DO NOT DISTURB sign on his door, and not permitting any more interruptions.

When interruptions can be shut out by a locked door or by a DO NOT DISTURB sign, this is the easiest way to avoid them. But some people—friends, parents, teachers, managers—have a responsibility to deal with unscheduled as well as scheduled requests for their time. Since they cannot refuse interruptions, they are left to schedule around them.

Their best defense is to do as Jean did, to devote time to organizing their tasks into subtasks of shorter duration, so that the devastating effects of interruptions are less costly.

The following observation can be helpful in estimating the length of time to complete a task that is subject to interruptions. (For background evidence for the observation refer to the results of Exercise 4.)

A task for which the required uninterrupted time U is the same as the expected time between interruptions ($i = 1/p$) will take an average of twice as long ($2U$) as it should because of the time wasted by interruptions.

This observation leads to the following rule-of-thumb for deciding whether to invest time to organize tasks into subtasks:

If you expect an average of i time units between interruptions, and if you value your time, then you should organize your tasks into subtasks each requiring a length of time substantially less than i.

Exercise 5.

Examine the results of Exercise 4. To what extent do they support the rule-of-thumb stated above? Under what circumstances would you be unwilling to follow the rule?

Exercise 6.

a) Suppose that an uninterruptible task is to be performed for which $U = 1/p$; show that

$$T = \left[U \left(\frac{U}{U-1}\right)^U - 1\right].$$

b) It follows from (**a**) that

$$\frac{T}{U} = \left[\left(\frac{U}{U-1}\right)^U - 1\right].$$

Collect evidence for the assertion that T/U is a decreasing function of U by completing the following table.

U	2	3	4	5	7	10	15	25	50	100	200
T/U											

c) For students with backgrounds in calculus. Apply l'Hospital's rule to show that

$$\lim_{U \to \infty}(T/U) = e - 1.$$

d) Above, we made the following observation:

> A task for which the required uninterrupted time is the same as the expected time between interruptions will take, on the average, about twice as long as it should, because of time wasted by interruptions.

Use the results obtained in (**b**) and (**c**) to refine this observation.

In Example 2, Jean, whose expected time between interruptions was 20 minutes, organized her assignment into five 15-minute subtasks. Her organizational method offered substantial time savings over Jim's "do-it-all-at-once" method. This was true even though Jean planned for a total of 75 minutes whereas Jim planned for only an hour.

The time savings that resulted from the extra time that Jean invested in organization illustrates a principle that any person who values time and who is beset with interruptions should seriously consider:

> Time invested in organizing a lengthy and complex task into shorter and simpler subtasks can be well rewarded by a reduction in the amount of time wasted by interruptions.

Although our examples and exercises encourage you to examine the effects of interruptions through calculations using Eq. (1), in real problems of your own interruptions may not be independent and devastating. You may be unable to supply good estimates for U and p. Thus you will be unable to use Eq. (1) and calculate T.

But even when you find the formula impossible to use, these ideas about interruptions can be useful in a general way. Your recognition that you may be able to shorten the overall time needed for certain tasks by organizing them into shorter subtasks can increase your control over your schedule and make you less vulnerable to the setbacks caused by interruptions.

Exercise 7.

Recall Gwynne, introduced at the beginning of the module. Gwynne is an upperclass resident advisor in a dormitory for college freshmen. Because of her responsibilities to that job, Gwynne's study schedule is often subject to interruptions. On a particular evening, Gwynne starts at 9 p.m. trying to complete a difficult mathematics problem that she estimates will take 15 minutes of uninterrupted concentration. Any interruption will require her to start over. At this time of the evening Gwynne estimates that her interruptions occur randomly and independently, and, on average, about once every ten minutes.

a) Under the stated conditions, how long can Gwynne expect to spend on her 15-minute problem?

b) Suppose that Gwynne is willing to spend five (uninterrupted) minutes organizing some of the problem information; as a result she will be able to complete the problem in three additional uninterrupted five-minute intervals. If interruptions occur with the pattern given above, how long will this scheme for solving her problem be expected to take?

c) Suppose that Gwynne resists organizing her solution method as described in (**b**). She asks, "Why should I deliberately choose to spend 20 minutes on a fifteen-minute problem?" Develop a response to her question.

Exercise 8.

My source for the following tale is the book by Herbert A. Simon entitled *The Sciences of the Artificial* (2nd Ed.), published in 1981 by MIT Press, Cambridge, MA. The development of stable subsystems is described therein as a critical factor in determining the speed of biological evolution or that of complex social systems.

Once upon a time, so the story goes, there were two watchmakers – Hora and Tempus. Each produced very fine watches that began to be in great demand. Their workshop phones rang frequently, bringing orders from new customers.

Both men made watches that consisted of about 1000 parts. Tempus had constructed his so that if several pieces were assembled and he had to put the assembly down – for example, to answer the phone – all the pieces fell apart and had to be reassembled from scratch. Naturally, the better his customers liked his watches, the more they told their friends – who interrupted his work to place orders for more watches. As a result it became increasingly difficult for Tempus to find enough uninterrupted time to finish a watch.

Hora, on the other hand, had designed his watches so that he could put together components of about ten parts. As with Tempus, if Hora was interrupted while working on a particular assembly the pieces fell apart and had to be reassembled from scratch. But the ten-part components, once completed, were "stable" and could be set aside for later use as a unit. Ten of these components could later be assembled into a larger stable subsystem. Eventually, the assembly of the ten larger sub-systems constituted the whole watch.

Hora prospered while Tempus became poorer and poorer and finally lost his shop. What was the reason?

To answer the question of the parable, we need additional numerical information. Let us use as a time unit the length of time required to add one part to the watch assembly. Then, to be specific, let us suppose that the probability of interruption during any given time unit is $p = 1/100 = 0.01$.

a) Under the stated conditions, how many time units will be needed for Tempus to complete a watch?

b) Under the stated conditions, how many time units will be needed for Hora to complete a watch?

c) If the time unit (the time required to add one watch part to the assembly) is five seconds long, how long can it be expected to take for Hora to complete a watch? For Tempus to complete a watch? Why did Hora propser and Tempus become poorer and poorer?

d) Traditionally, a story called a parable is accompanied by a moral. What moral can be learned from this parable?

Exercise 9.

Jane Jacobs is an assistant vice president for a large financial institution. In addition to coping with the day-to-day responsibilities of her current position, Jane is anxious to show that she is capable of a higher level of leadership. She believes that she has good ideas about the future growth of her institution. But, for her input to receive the serious consideration of the president and of Jane's associates, Jane's ideas must be well thought out and presented in a written proposal. Jane has been keeping notes of her ideas but estimates that she needs an uninterrupted interval of two hours to write the proposal. Interruptions—matters needing her immediate attention—come on the average of once every half hour.

Instead of using minutes as the unit with which she measures time, Jane measures her time in quarter hours. "In less time than that, I can't get a thing done," she comments. Her report will require $U = 8$ consecutive uninterrupted quarter hours. In a given quarter hour she has probability $p = 1/2 = 0.5$ of interruption. Interruptions for Jane are independent and devastating. When an interruption occurs, it completely diverts her thinking so that she must rethink her reasoning up to the point of the interruption when she returns to work on her report.

a) Using the information given above and in Eq. (1), estimate the number of working hours it will take Jane to complete her report.

b) The result from (**a**) might be described as "ridiculous." We can hardly expect Jane to invest more than 100 hours in writing her report. After a while, we would expect her either to give up or to give her secretary orders to prevent all interruptions until the report gets finished.

However, avoiding interruptions may be undesirable. Jane's so-called interruptions are part of her job responsibility. (Our use of the term "interruptions" designates events that are unscheduled—with no reference to whether or not they are desirable or necessary.)

There are other alternatives that Jane could consider. She could structure her report writing in a different way. Suppose that, after thinking about the problem, Jane estimates that she could complete her report in six half-hour intervals of time. She would use the first half-hour to organize her notes into an outline with four main parts. The next four half-hours would be used to write the four main parts of the proposal. The final half-hour would be used to edit and synthesize the four parts into a unified proposal. However, Jane resists this alternative. "This would require me to devote three hours to a task that should take only two hours," she protests.

Calculate an estimate of the length of time it will take Jane to complete her proposal using this scheme of organization.

c) Based on the results calculated in (**a**) and (**b**) and on calculations based on time-organization schemes you invent, develop a recommendation for Jane. What do you advise? Explain why.

(**d**) Reconsider (**b**) using minutes as the time unit. In this case $U = 30$ and $p = 1/30$. Why is the calculated length of time so much less in this case?

Exercise 10.

For this exercise you will find it useful to refer to the tables of (**a**), (**b**), and (**c**) in Exercise 4.

Calculate the amount of time that could be saved by reorganizing an uninterruptible 20-minute task: (**i**) into two ten-minute sub-tasks; (**ii**) into four five-minute sub-tasks when the probability of interruption during any given minute is (**a**) $p = 0.01$; (**b**) $p = 0.05$; (**c**) $p = 0.10$; (**d**) $p = 0.20$; (**e**) $p = 0.25$.

f) If the reorganization of a 20-minute task into two ten-minute uninterruptible sub-tasks will require extra time of five uninterrupted minutes just to get organized, in which of the situations (**a**)-(**e**) would this extra time for organizing be worthwhile?

g) If the reorganization of a 20-minute task into four five-minute uninterruptible sub-tasks will require two extra uninterrupted five-minute intervals just to get organized, in which of the situations (**a**)-(**e**) would the extra organizational time be a worthwhile investment?

Exercise 11.

Perhaps you know someone like Fred Biddlesmith who is on academic probation at a nearby college. Last term, as a first semester freshman, Fred did not do well in his courses. In particular, he failed his introductory statistics course. Fred is a bright young man who is capable of college work. Nevertheless, he has had difficulty completing the tasks he starts. With our previous discussions and examples in mind, invent a description of what may have been Fred's situation. How might he remedy things for the future?

Exercise 12.

Kim McNulty was a journalism major in college and wrote for a variety of student publications. She also helped the college public relations office, providing text and layout for college publications. Now Kim is married and staying at home caring for two young children. She would like to do freelance writing but says, "I never have blocks of time large enough to get anything done." Develop some suggestions for Kim.

Exercise 13.

Consider the activities of faculty members Kline and Lewis last Tuesday morning. Both were in their offices preparing materials for their 11 a.m. classes.

Professor Kline's door was open and, as she worked at her desk, students dropped in to chat and to ask questions about homework assignments. Feeling obliged to help them, Professor Kline pushed aside her preparations and dealt with their questions. The students went away satisfied. At 11 a.m., Professor Kline conducted a disorganized, poorly-prepared class.

Professor Lewis, on the other hand, spent Tuesday morning with the office door closed. Instead of responding to knocks on his door, he diligently prepared his lesson. Some of his students were disgruntled about their professor's unavailability. But Professor Lewis' 11 a.m. class was well-organized and effective.

a) Discuss the conflicting responsibilities that Professors Kline and Lewis face.

b) Formulate suggestions for each of these professors about how to meet their conflicting responsibilities.

Exercise 14.

It is a common belief that since interruptions are unscheduled events one cannot plan for them. Develop a convincing argument that this is not so. Argue that even though interruptions are unscheduled events one can make plans that allow for them and minimize their effect.

Exercise 15.

Make a list of tasks for which interruptions cause serious setbacks to your efforts. For which of these tasks could you reduce the overall time required if you spent time organizing the task into shorter sub-tasks?

Exercise 16.

Focus on an interruption problem (either real or hypothetical) that interests you. Describe the problem in detail; assess whether the interruptions are independent and devastating; estimate values for U and p and apply Eq. (1). Consider ways to divide the task into sub-tasks; attempt to discover a best way of organizing and completing the task.

5. Derivation of Interruption-time Formula

Suppose that a certain project is to be undertaken and that its completion requires U uninterrupted time units.

Further suppose that interruptions occur randomly, with a specified probability p during any given time unit. As in our examples, we shall restrict our attention to situations for which interruptions are independent and devastating. That is, the presence or absence of an interruption during one time unit has no effect on the probability of subsequent interruptions, and each interruption requires restarting the project.

Our goal is to derive a formula for T, the average number of time units that will be required to complete the desired project while subject to interruptions. Our derivation relies on three facts from the theory of probability:

1) If the probability that an event will occur in a given time unit is p, then the probability that it will not occur is $1 - p$.

   For example: If the probability that an interruption will occur during any specified minute is $p = 0.1$, then the probability of no interruption during that minute is $1 - p = 0.9$.

2) For a collection of independent events, the probability that all will occur is the product of their individual probabilities.

   For example: If the probability of interruption during any specified minute is $p = 0.1$, then the probability of interruptions occurring in each of two consecutive minutes is $(0.1)(0.1) = .01$. Similarly, the probability of two consecutive minutes being free from interruptions is $(0.9)(0.9) = 0.81$. The passage of three consecutive minutes without interruption has probability $(0.9)(0.9)(0.9) = 0.729$, and so on.

3) If one specified outcome out of a group of possible outcomes has probability p, then p estimates the proportion of times that the specified outcome will occur in the long run. Furthermore, $1/p$ estimates the expected or average number of repetitions needed to produce the specified outcome once.

   For example: When the probability of interruption during any given minute is $p = 0.1$, then in an 8-hour work day (480 minutes) the expected proportion of minutes that are interrupted is $p = 0.1 = 10\%$ and the estimated number of interruptions per day is 48. On average, the number of minutes that will need to pass to produce one interruption is $1/p = 1/(0.1) = 10$.

We now proceed to derive the formula

$$T = \frac{1}{p} \left[\frac{1}{(1 - p)^{U}} - 1 \right],$$

where T is the expected number of time units required to complete the desired project while subject to interruptions. The derivation rests on substitution of the correct expressions into the following verbal equation:

$$\begin{pmatrix}\text{Average number of}\\ \text{time units required}\\ \text{to complete project}\end{pmatrix} = \begin{pmatrix}\text{Average number of}\\ \text{time units lapsed}\\ \text{per interruption}\end{pmatrix} \cdot \begin{pmatrix}\text{Average number of}\\ \text{interruptions}\\ \text{per project}\end{pmatrix}$$

The left-hand side of the verbal equation designates T. From item (3), above, we see that the first factor on the right-hand side is $1/p$. It remains for us to determine the second factor.

Since the probability of an interruption during any given time unit is p, the probability that a time unit will pass without interruption is $1 - p$. Applying item (2), above, the probability, that U consecutive time units will pass without interruption is $(1 - p)^U$. Using item (3), we concluded that if the project is attempted a large number of times, the proportion of starts completed will be $(1 - p)^U$. Continuing to apply the information from item (3), we observe that $1/(1 - p)^U$ is the expected (or average) number of starts required to produce an interruption. Because one start (the final, successful one) necessarily involves no interruption, the average number of interruptions is one less than the average number of starts.

Thus the average number of interruptions per project is

$$\left[\frac{1}{(1 - p)^U} - 1\right] .$$

Substitution of this quantity for the second factor in the verbal equation completes the derivation of Eq. (1).

6. Scheduling Around Interruptions

Because interruption patterns are likely to vary, successful scheduling around interruptions may be more of an art than a science. Even so, experience with stereotypical problems can develop an individual's awareness of the posssible scheduling options that are available and of how well they might work.

To investigate further how to control the effects of interruptions, we consider a job-related example.

Example 3:

Kurt works in software development for a large architectural firm. His work requires periods of uninterrupted concentration. However, as a senior member of his work group, Kurt is also responsible for coordinating software development and for helping his co-workers solve some of the problems that they encounter. These responsibilities lead to interruptions. For Kurt, the probability of interruption in a given hour is $p = 0.3$. How should he organize his work to get the most done in a four-hour morning?

Analysis:

To begin, we note that Kurt's interruptions have not been described as independent and devastating. In fact, it is quite likely that they are not independent; for example, if Kurt deals with many interruptions between 8 a.m. and 9 a.m., this may reduce his chances of interruption between 9 a.m. and noon. However, since we have little information about Kurt's situation, we will suppose the worst (i.e., the situation in which he has least control). We suppose that his interruptions are both independent and devastating and that their probability cannot be reduced. For situations in which Kurt has some control over the pattern and effects of interruptions, the results we obtain will be useful as a worst case against which various schedule improvements can be evaluated.

Kurt's probability of interruption in any given hour has been specified as $p = 0.3$. Using Eq. (1) we can estimate that interruptions will extend the length of an uninterruptible task of length $U = 1$ hour to

$$T = 1.43 \text{ hours.} \qquad \text{(A)}$$

But consider an alternative formulation of the same one-hour task. Suppose a *minute* is used as the time unit. The probability of interruption during any given minute is $p = (0.3)/60 = 0.005$. With $U = 60$ minutes and $p = 0.005$, Eq. (1) yields

$$T = 70.2 \text{ minutes } (1.17 \text{ hours}). \qquad \text{(B)}$$

(Note: An individual whose responsibilities permit control of interruptions—such as dealing with them all between 8 a.m. and 9 a.m. and allowing 9 a.m. to noon for uninterrupted tasks—is encouraged to consider this option. A schedule that combines interruptions and brief tasks in one block of time, and reserves other blocks of time free from interruptions, offers an opportunity for high efficiency.)

Why do the values (A) and (B) differ?

The difference between value (A) and (B) depends on an aspect of Eq. (1) that our previous discussion has not emphasized. See, however, Exercise 9(d). Equation (1) measures the effect of an interruption in this way:

> When an interruption takes place, it eliminates productive value for *the entire time unit* in which it occurs.

Thus, if time is expressed in hours, Eq. (1) treats each interruption as if it eliminates an entire hour of time-on-task. On the other hand, if time is expressed in minutes, Eq. (1) calculates time lapsed as if each interruption has eliminated only a minute of time-on-task.

Continuing with our analysis of Kurt's morning at work, let us suppose that he estimates that interruptions require an average of ten minutes of time. Based on his estimate, we decide that a ten-minute block is a suitable unit of time. In this case, an hour project will require $U = 6$ ten-minute blocks and the probability of interruption during any ten-minute block will be $p = (0.3)/6 = 0.05$. Using Eq. (1) we obtain

$T = 7.2$ ten-minute blocks.

To continue with our investigation of how Kurt should spend his morning, we will do some experimental calculations. Table 1 includes a variety of lengths of tasks together with estimates of the time required for their completion. In each case, the probability of interruption during any ten-minute time interval is $p = 0.05$.

Uninterrupted time U required for task	Total time T that task will require because of setbacks caused by interruptions
24 ten-minute intervals (4 hours)	48.5 ten-minute intervals (8.1 hours)
18 ten-minute intervals (3 hours)	30.4 ten-minute intervals (5.1 hours)
15 ten-minute intervals (2½ hours)	23.2 ten-minute intervals (3.9 hours)
12 ten-minute intervals (2 hours)	17.0 ten-minute intervals (2.8 hours)
9 ten-minute intervals (1½ hours)	11.7 ten-minute intervals (1.95 hours)
6 ten-minute intervals (1 hour)	7.21 ten-minute intervals (1.2 hours)
5 ten-minute intervals (50 minutes)	5.85 ten-minute intervals (58.5 minutes)
4 ten-minute intervals (40 minutes)	4.55 ten-minute intervals (45.5 minutes)
3 ten-minute intervals (30 minutes)	3.33 ten-minute intervals (33.3 minutes)
2 ten-minute intervals (20 minutes)	2.16 ten-minute intervals (21.6 minutes)
1 ten-minute interval (10 minutes)	1.05 ten-minute intervals (10.5 minutes)

Table 1. Time estimates for various lengths of tasks for Kurt.

Examination of Table 1 leads to the following observations:

On the average, in a four-hour morning, Kurt will not finish a task that needs more than 2.5 hours of uninterrupted time.

The shorter the tasks that Kurt attempts, the less time he wastes on interruptions.

We cannot formulate good advice for Kurt unless we have additional information about the nature of the tasks he must complete. However, we can consider some hypothetical situations. Suppose that a typical task for Kurt requires 1.5 uninterrupted hours. From Table 1 we see that actual completion of such a task would require 117 minutes on the average. Might Kurt save time overall by taking some time to organize the tasks into three 30-minute sub-tasks? Three such sub-tasks will take about 100 minutes altogether. (Refer to Table 1; $(3)\times(33.3)=100$.) If Kurt's organizing time is less than 17 minutes, organizing will save time; otherwise it will be quicker to cope with setbacks caused by interruptions.

What should Kurt do about an uninterruptible three-hour job assignment? If he insists on being punctual for his noon-time cafeteria appointment, then the expected 5.1-hour duration for such a task makes its completion appear almost impossible, unless it is restructured. Suppose Kurt can, in an extra uninterrupted hour, reorganize the three-hour task into three one-hour tasks. The total time required altogether will be almost five hours ((4) (1.2 hours) = 4.8 hours). Despite the fact that the reorganization saves little time it is an important strategy in this case. Its advantage is that it changes the task from one that would be difficult to ever get done in a four-hour work interval into one whose sub-tasks can be readily accomplished—in two work mornings, if not in one.

Kurt's problem illustrates some of the variety of possibilities available for scheduling around interruptions. Although you may be reluctant to apply such tedious analysis to your own scheduling problems, you can translate your work with Kurt's schedule possibilities into general rules-of-thumb that can help you develop work schedules that lead to tasks accomplished instead of frustration.

If you think about it first, doing it will be easier.

7. More Exercises

17. a) Suppose Kurt's probability of interruption during any given hour is $p = 0.6$. Develop a table with the format of Table 1 with estimates of the times that tasks of various uninterruptible lengths can be expected to take.

b) Suppose Kurt has received assignments for next week. A quick scan of his list of tasks has led him to size them up as follows:

one 4-hour uninterruptible task;
two 2-hour uninterruptible tasks;
four 1-hour uninterruptible tasks.

Suppose also that for Kurt to divide a task into four uninterruptible subtasks of equal size requires an investment of organizational time that is 25% as long as the task being organized. If Kurt is unwilling to subdivide a task (into four parts) more than once, can he expect to complete all of these tasks in five four-hour mornings? If so, provide a possible schedule for him.

18. A precision product requires deft tooling by a skilled worker. If she is interrupted during the tooling process, the product will be unacceptable and a startover will be required. (One cause of interruption in this case may be inattention resulting from fatigue. For this kind of interruption the property of independence would probably not be satisfied.) So that Eq. (1) may be used, we restrict our attention to situations in which interruptions are independent.

a) If the delicate tooling operation requires ten minutes and if the probability of interruption during any given minute is $p = 0.1$, how many minutes, on average, are needed to produce an acceptable product?

b) If the single delicate operation described in (**a**) could be replaced by a pair of tasks that each require six uninterrupted minutes, on the average, how many minutes would be needed to produce an acceptable product?

c) Using (**a**) or (**b**), whichever is more efficient, how many precision products could be made by the skilled worker in an eight-hour day?

d) Reconsider (**a**), (**b**) and (**c**) if $p = 0.05$.

e) If you owned the company that produces the precision product, what would you do if presented with results of the type calculated in (**a**)-(**d**) above? Explain your decision.

19. *Test yourself.*

a) Invent a situation similar to that of Kurt in Example 3 or of the skilled worker in Exercise 18. Carefully describe the details of the situation. Estimate values for U and p. Under what circumstances are the interruptions likely to be devastating and independent? Develop hypothetical schedules for completing the desired tasks and compare the total times that the different schedules require. In the situation you describe, are interruptions a serious enough problem to warrant the investment of extra time for organization?

b) Think of a situation in which you have had trouble completing a task because of interruptions. Apply the ideas presented in this module to develop strategies for reducing the overall time required by the task.

8. Hints and Answers for Exercises

1. The mean length of time for the 50 experiments is $523/50 = 10.5$ minutes.

2. Accurate guesses require experience with time assessment, but your guesses should have the following relationship:

Max's time > Marsha's time > Melanie's time

a) For Max, $T = 41.6$ minutes.

b) For Marsha, $T = 18.7$ minutes.

c) For Melanie, $T = 13.4$ minutes.

d) Max's task requires a longer interval of time than the average time between his interruptions. Thus most of his starts result in effort wasted; only rarely does his interruption pattern allow the time he needs.

3. a) T = 15.4 hours.

b) T = 104.5 minutes = 1¾ hours.

c) If Mike and Michelle are serious about card-building, they should make arrangements for Mitch to be elsewhere.

4. Some of the values of T for Tables of (**a**), (**b**), and (**c**) in Exercise 4 should not be taken seriously; in actual situations, human frustration will take over and prevent their occurrence!

a) 5.2 min, 5.8 min, 6.9 min, 10.3 min, 12.9 min, 19.8 min, 62 min.

b) 10.6 min, 13.4 min, 18.7 min, 41.6 min, 67.0 min, 170 min, over 2000 min.

c) 22.3 min, 35.8 min, 72.3 min, over 400 min, more than 1000 min, almost 10,000 min, more than 2 million min.

d) As p increases, T increases.

e.) No, no, yes, yes; no, no, yes.

f) The observation is "roughly" true. More precisely:

in Table (b), T = 1.87U; when U = i = 10
in Table (c), T = 1.79U; when U = i = 20

Intuitively we might expect that if our interruptions occur on an average of every i minutes, then we would usually be able to accomplish tasks of length i. However, because the interruptions are random, they will not necessarily be spaced at intervals that will permit us to accomplish the whole task. In fact, sometimes we will have almost completed the task when an interruption occurs and requires us to start over. The random nature of the interruptions causes the possibility that a great deal of time will be wasted through starting over.

g) If p = 0.25, three five-minute tasks will take 38.7 minutes; one ten-minute task will take 67 minutes.

h) If p = 0.1, three ten-minute tasks will take 56.1 minutes; one 20-minute task will take 72.3 minutes.

i) When U is considerably larger than the expected time $i = 1/p$ between interruptions, an uninterrupted time interval of length U is rare. In such a situation, the time required to complete a task that needs U uninterrupted time units becomes excessive because of time wasted on start-overs caused by the random interruptions.

Your calculations should give you evidence to conclude that a task of length $U > 1/p$ may take longer than a group of shorter tasks whose combined length is greater than U.

5. Answers to parts g, h, and i for Exercise 4 offer evidence in support of the rule-of-thumb. It is not, however, always rewarding to subdivide a task into subtasks: sometimes the time required to organize takes longer than the time saved; and sometimes the work on minor tasks begins to seem mindless and petty and—even though easier and quicker—not as much fun as working on the project as a whole.

6. a) In Equation (1) substitute for p the fraction $1/U$.

b) Value of U 2, 3, 4, 5, 7, 10, 15, 25, 50, 100, 200

Value of T/U 3, 2.375, 2.160, 2.052, 1.942, 1.868, 1.815, 1.775, 1.746, 1.732, 1.725

c) Let $U = 1/h$ and find the limit as $h \to 0$.

d) The given observation may be refined to read: A task for which the required uninterrupted time is the same as the expected time between interruptions will take, on the average, about $e - 1$ (1.718) times as long as it should, because of time wasted on startovers due to interruptions.

7. a) About 39 minutes.

b) About 28 minutes.

c) The answers to (**a**) and (**b**) may be sufficient to convince Gwynne. However, if she fails to accept the results of our calculations because she does not understand the procedure we have used to obtain them, then it may be useful to respond to her objection in the following way:

Your investment of five extra minutes is not an investment of extra time in your mathematics problem but a device for coping with the effects of the interruptions that are part of your job. The time wasted (39 − 15 = 24 minutes) because of startovers caused by interruptions can be reduced considerably by investing time to get organized.

8. a) FOREVER! (Calculation using the formula for *T* gives a value in excess of 2 million time units.)

b) Hora must assemble a total of 111 ten-part components; each takes an average of 10.57 time units. Thus, assembly of an entire watch will take about 1170 time units, on the average.

c) For Hora, assembly of a watch will take about 1.6 hours. For Tempus, calculations predict a time that exceeds 3000 hours—but can we believe them? Perhaps Tempus's downfall came because he was a stubborn fellow who said, "Why should I invest extra time to organize my procedure into small steps? Already it takes too long to make a watch, and organizing would make it take even longer!"

d) MORAL (paraphrasing Christopher Robin): Organization is what you should do before you do something so that when you do it, you get it done.

9. a) The formula for *T* gives an estimated time of 127.5 hours. Is this a believable answer?

b) Each half-hour task will take 1.5 hours, allowing for the effects of interruptions. Jane's total time thus will be nine hours, STILL TOO LONG.

c) Jane's interruption pattern prevents her from completing all but very short uninterruptible tasks. A possible alternative for her is one that many busy executives use: work after hours—at night or on weekends—when long intervals of uninterrupted time can be found.

d) In this case, the time required for Jane to accomplish one half-hour segment of her task is $T = 53$ minutes. The total proposal (six half-hour segments) will thus take about 318 minutes, or five hours and 18 minutes. This calculated value is substantially less than the time of 9 hours calculated in (**b**) because of the different choice of time unit. In (**b**) the use of a 15-minute block as the time treated each interruption as if it wasted the entire 15-minute block in which it occurred. In this latter case, however, each interruption is treated as if it wastes only the minute in which it occurs.

The accuracy of the predictions of Eq. (1) depends on the choice of time unit; sound results require using a time unit whose length agrees with the expected length of interruptions.

10. **a**) (**i**) about 1 minute; (**ii**) about 1.5 minutes.
b) (**i**) about 9 minutes; (**ii**) about 13 minutes.
c) (**i**) about 35 minutes; (**ii**) about 45 minutes.
d) (**i**) about 345 minutes; (**ii**) about 388 minutes.
e) (**i**) over 1100 minutes; (**ii**) over 1200 minutes.
f) In (**a**) the suggested reorganization would require an average of 5.2 minutes and would not be worthwhile. In (**b**) the reorganization would require about 5.8 minutes; the savings of 3.2 minutes would be especially worthwhile if it could be repeated many times, by reorganizing many tasks into shorter sub-tasks. In (**c**), (**d**), and (**e**) organizing saves time. However, it is reasonable to ask whether even reorganization reduces the task-time to a tolerable length.
g) In (**a**) the suggested reorganization would require an average of 10.4 minutes and would not be worthwhile. In (**b**) the reorganization would require, on the average, about 11.6 minutes and the time savings of just over one minute would be marginally worthwhile. In (**c**) the investment of organizational time is definitely worthwhile and even offers a savings of about three minutes over the reorganization suggested in (**f**). In (**d**) and (**e**) this reorganization offers a considerable time savings from the original and from the reorganization suggested in (**f**). It is important, however, to evaluate time cost as well as time savings; is the task worth it? Is the task one that justifies locking the door on interruptions?

11. One possibility: Fred has a short attention span. Distractions tend to draw his mind away from his studies after about 15 minutes of work. If Fred recognizes and accepts this characteristic of himself, he can learn to work around it. For example, he can begin assignments by skimming his reading and class notes and by making a list or outline of important concepts. Using the outline, he can proceed to fill in details by working in short intervals. Fred can be an effective student in spite of his short attention span if he is willing to take time to organize his studies.

12. If Kim genuinely wishes to get some writing projects done, then the demands of her at-home schedule require her to begin by organizing. Her first step could be the investment of time for developing outlines or plans for the articles she would like to write. Once she has an outline clearly in mind, she can work on various sections of it in the short intervals of time that she finds available.

13. a) Professors Kline and Lewis are responsible both for finding time to prepare effective classes and for finding time to respond to the needs of individual students. Not surprisingly, they find it impossible to do both at the same time.
    b) Because many of a professor's duties require uninterrupted intervals of time, and because the adequate completion of these (lesson preparation, paper grading, research) is as important to students as the time for individual meetings with them, it seems reasonable for a professor to divide out-of-class time into two categories: some time should be scheduled as available for interruption (the professor's "office hours"); other times may be scheduled as unavailable for interruption. When the professor is available for interruption, the free time between interruptions can be used to accomplish brief tasks like reading mail, drafting memos, and completing brief sub-tasks identified by organizing lengthy "uninterruptible" tasks.

14. This one's for YOU! Are you convinced that organized planning can reduce the effects of interruptions? Explain why and how.

15. In making your list, consider internally-caused interruptions (such as fatigue, lack of interest, short attention span) as well as externally-caused ones (such as visitors and noise).

    After you make your list consider the following: Many people dislike the thought of investing time before they begin a project to get organized; is this true for you? Is this a costly attitude to have? Is it ever beneficial?

16. It may be hard to fit a problem of your own into the framework we have discussed. You may, for example, have to guess at numerical values. Even if your values for U and p are simply guesses, working through the problem can be a useful experiment. At the end of your effort, one possible conclusion is that the analysis has little value or provides little insight into the problem with which you began. The solution methods we learn do not always apply to the problems we encounter; a sensible procedure is to try a method if we think it might work, but to try it as an experiment whose results may need to be rejected.

17. a) The probability that Kurt will be interrupted in any given ten-minute interval is $p = 0.6/6 = 0.1$. The times that various tasks will require are summarized in the table below.

Uninterrupted time U required for task	Total time T that task will require because of setbacks caused by interruptions
24 ten-minute intervals (4 hours)	115 ten-minute intervals (19.2 hours)
18 ten-minute intervals (3 hours)	55.6 ten-minute intervals (9.4 hours)
15 ten-minute intervals (2½ hours)	38.6 ten-minute intervals (6.4 hours)
12 ten-minute intervals (2 hours)	25.4 ten-minute intervals (4.2 hours)
9 ten-minute intervals (1½ hours)	15.8 ten-minute intervals (2.6 hours)
6 ten-minute intervals (1 hour)	8.8 ten-minute intervals (1.5 hours)
5 ten-minute intervals (50 minutes)	6.9 ten-minute intervals (69 minutes)
4 ten-minute intervals (40 minutes)	5.24 ten-minute intervals (52 minutes)
3 ten-minute intervals (30 minutes)	3.72 ten-minute intervals (37 minutes)
2 ten-minute intervals (20 minutes)	2.35 ten-minute intervals (23 minutes)
1.5 ten-minute intervals (15 minutes)	1.71 ten-minute intervals (17 minutes
1 ten-minute interval (10 minutes)	1.11 ten-minute intervals (11 minutes)

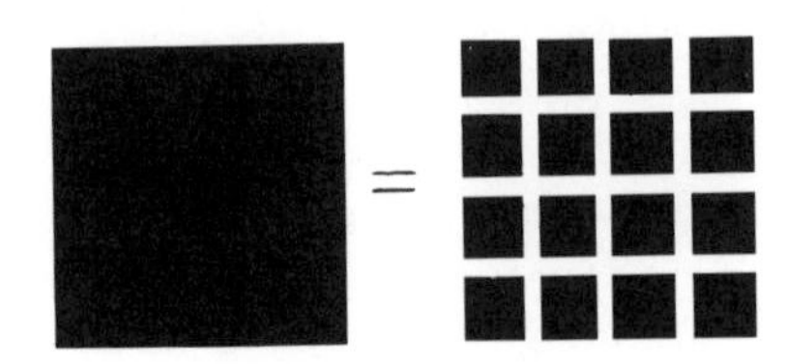

b) From the table above, we see that Kurt must reorganize his four-hour and two-hour tasks into sub-tasks if he wishes to get them done within four-hour work intervals. His single four-hour task can be subdivided and completed in five 88-minute time intervals. His two-hour tasks can both be completed in a total of ten 37-minute intervals. Each one-hour task, if reorganized will require five 17-minute intervals. Twenty such intervals will be required altogether. The total time of 85 minutes ($5 \times 17 = 85$) for the reorganized one-hour task does not offer much savings over the 88 minutes Kurt would expect it to take if undivided, however, subdividing this task offers schedule flexibility since one or two 17-minute tasks can be squeezed in where an 88-minute task won't fit.

One possible schedule for Kurt would have him tackle one sub-task from each of his seven assigned tasks during each morning. If he does this, then out of 240 available minutes he is scheduling a total of

$$88 + 37 + 37 + 17 + 17 + 17 + 17 = 230 \text{ minutes.}$$

Will this schedule work? In part this depends on the accuracy of Kurt's original time estimates. Also, because our calculated values are averages rather than precise times, the schedule may give Kurt some mornings with lots of extra time and some mornings with too much to do. Part of the art of scheduling involves learning how much slack time is needed in a schedule so that bad time estimates do not lead to too much frustration.

18. When one is tired, mistakes tend to cause frustration that leads to additional mistakes. When interruptions cause additional interruptions instead of occurring in a random pattern, they are not independent.

a) 18.7 minutes

b) 17.6 minutes

c) If a worker could continue to work at this tooling operation for eight consecutive hours, an average of 27 precision products could be produced.

d) 13.4 minutes, 14.4 minutes; 33 precision products

e) First, I would check to see how many parts are actually being produced. If the number is low, I would try to discover why and find a remedy. Any suggestions?

19. You're on your own for this one!

Don't be timid about wild guesses at numerical values for U and p. When you calculate using Eq. (1), if your answers seem unreasonable you can use them as a guide to devise more reasonable starting values.

Don't be hesitant to criticize the results you obtain when you apply this module's ideas to your own situation. Discriminating use of knowledge has two aspects: recognizing when a concept is useful and recognizing when it is not.

UMAP

Modules in Undergraduate Mathematics and its Applications

Module 658

Windchill

William Bosch
and
L. G. Cobb

Published in cooperation with the Society for Industrial and Applied Mathematics, the Mathematical Association of America, the National Council of Teachers of Mathematics, the American Mathematical Association of Two-Year Colleges, and The Institute of Management Sciences.

COMAP

INTERMODULAR DESCRIPTION SHEET: UMAP Unit 658

TITLE: WINDCHILL

AUTHORS: William Bosch
Department of Mathematics and Applied Statistics
University of Northern Colorado
Greeley, Colorado 80639
and
L. G. Cobb
Department of Earth Science
University of Northern Colorado
Greeley, Colorado 80639

CLASSIFICATION: Precalculus

APPLICATION FIELD: Meteorology

ABSTRACT: This module develops formulas for heat loss and wind chill in both metric and English units.

PREREQUISITES: a) intermediate algebra skills, b) completing the square, c) graphing quadratic functions.

Windchill

by

William Bosch
Department of Mathematics and Applied Statistics
University of Northern Colorado
Greeley, Colorado 80639

and

L. G. Cobb
Department of Earth Science
University of Northern Colorado
Greeley, Colorado 80639

Table of Contents

INTRODUCTION . . . 1
1. THE HEAT LOSS FORMULA . . . 1
2. DISCUSSION OF THE HEAT LOSS FORMULA . . . 2
3. THE LIMITING EFFECT OF WIND SPEED . . . 3
4. CONVERTING TO OTHER UNITS OF MEASUREMENT . . . 5
5. THE WINDCHILL INDEX . . . 6
6. BIBLIOGRAPHY . . . 10
7. MODEL EXAM . . . 10
8. ANSWERS TO EXERCISES . . . 11
9. ANSWERS TO MODEL EXAM . . . 14

Modules and Monographs in Undergraduate Mathematics and its Applications Project (UMAP)

The goal of UMAP was to develop, through a community of users and developers, a system of instructional modules in undergraduate mathematics and its applications to be used to supplement existing courses and from which complete courses may eventually be built.

The Project was guided by a National Advisory Board of mathematicians, scientists, and educators. UMAP was funded by a grant from the National Science Foundation and is now supported by the Consortium for Mathematics and Its Applications, Inc. (COMAP), a nonprofit corporation engaged in research and development in mathematics education.

This material was prepared with the partial support of National Science Foundation Grant No. SED80-07731. Recommendations expressed are those of the authors and do not necessarily reflect the views of the NSF or the copyright holder.

Introduction

"How cold is it?"

"Twenty-five degrees."

"It sure feels colder than twenty-five degrees! I rushed out this morning without my cap and gloves; so when I walked between buildings on campus a few minutes ago, I felt as if I had frostbite on both ears and fingers. That wind is something else!"

Air motion, or wind, has a significant impact on the temperature that a person feels, often making it seem colder than it actually is. Under certain circumstances, what is "felt" can be a more important indicator of the effects of temperature than temperature itself. Thus, attempts have been made to quantify the effect that wind has on the temperature one feels.

Research into the problem of how much heat a body will lose under given conditions of temperature and wind has resulted in a "Windchill Index" that represents the equivalent temperature that exposed skin would feel with little or no wind. Exposure to cold temperatures associated with windy conditions can be dangerous if proper precautions are not taken. Explorers of polar regions and seamen were some of the first people who needed to assess such risks. More recently, farmers, firemen, and others involved in outdoor activity have become concerned about the chilling effect of the wind. So weather reports on radio and television commonly include information about "windchill" if it is a cold, windy day.

How "chilly" one feels depends upon a number of things besides wind and temperature, such as state of nourishment, individual body metabolism, and the moisture content of the air. However, the Windchill Index is a good guide to the type and amount of clothing that will be needed for given conditions of wind and temperature and is used widely for such purposes.

1. The Heat Loss Formula

Paul A. Siple, who accompanied Admiral Byrd to the Antarctic, and Charles F. Passel were among the first researchers to measure the chilling effect of winds at low temperatures (Siple and Passel, 1945). The Quartermaster Corps and the Medical Research Laboratory of the U.S. Army continued the research during and after World War II. Siple and Passel conducted experiments in Antarctica about 1940 and developed a heat loss formula by measuring the rate of freezing of water in a cylindrical bulb at various temperatures and wind speeds. By fitting a curve to the observed values, they obtained the following formula:

$$H = (10.45 + 10\sqrt{v} - v)(33 - T) \tag{1}$$

where

v is the speed of the wind in meters per second (m/s)
T is the air temperature in degrees Celsius (°C), and
H is the heat loss in kilogram calories per square meter of surface area per hour ($\text{Kcal}/m^2/\text{hr}$).

2. Discussion of the Heat Loss Formula

The normal mean temperature of the human body is 37°C (98.6°F); the skin, however, is cooler. Human skin, when protected by adequate clothing in cool weather, maintains a temperature of approximately 33°C. Thus, the factor $(33 - T)$ in Eq. (1) is simply the difference between skin temperature and air temperature. Equation (1) assumes that the air temperature does not exceed 33°C; otherwise, the body would be receiving heat from the air rather than losing it to the air. Siple and Passel assumed that heat loss for a constant wind speed would be directly proportional to this difference. The factor $(10.45 + 10\sqrt{v} - v)$ in Eq. (1) was obtained by fitting a curve to observed values.

Equation (1) indicates the amount of heat lost rather than the equivalent temperature felt by the body. The relation between the two will be discussed in Section 5.

Exercises

1. By conducting experiments with human subjects, Siple and Passel found that exposed areas of an individual's face will freeze within 30 seconds (on the average) if heat loss equals or exceeds 2300 kilogram calories per square meter per hour.

   (a) At what temperature will this heat loss be attained if the wind speed is 4 meters per second?

   (b) At what temperature will this heat loss be attained if the wind speed is 20 meters per second?

2. Siple and Passel also found that exposed areas of an individual's face will freeze within one minute (on the average) if heat loss equals or exceeds 2000 kilogram calories per square meter per hour.

   (a) At what temperature will this heat loss be attained if the wind speed is 4 meters per second?

   (b) At what temperature will this heat loss be attained if the wind speed is 10 meters per second?

3. The Limiting Effect of Wind Speed

In Eq. (1), the heat loss due to wind speed is embodied in the factor $(10.45 + 10\sqrt{v} - v)$, which we denote by L:

$$L = (10.45 + 10\sqrt{v} - v). \tag{2}$$

If we regard the right side of (2) as a quadratic function of $\sqrt{v}$, we can complete the square as follows:

$$\begin{aligned} L &= -(v - 10\sqrt{v} \qquad) + 10.45 \\ &= -(v - 10\sqrt{v} + 25) + 35.45 \\ &= -(\sqrt{v} - 5)^2 + 35.45 \end{aligned}$$

Since the square of a real number is never negative, we see that L has a maximum value of 35.45 and that this occurs when $v = 25$ m/s. (See Figure 1 for a graph of L as a function of v.)

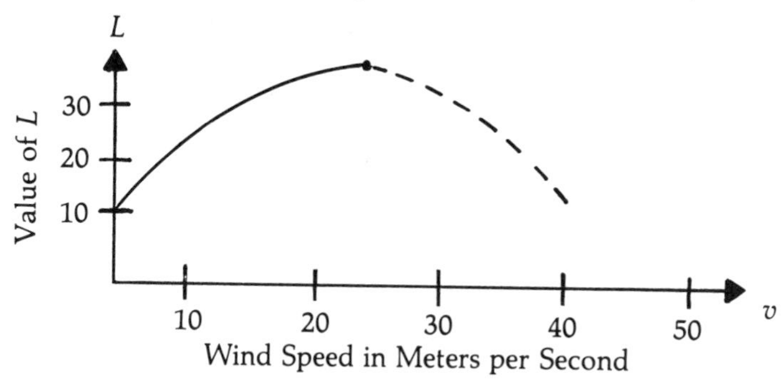

Figure 1. The maximum heat loss due to wind occurs at a velocity of 25 meters per second.

Figure 1 indicates that winds in excess of 25 m/s will have a less chilling effect than a wind of 25 m/s. This may be the case or it may be a defect of the model. A little reasoning suggests that the chilling effect does not diminish for wind speeds over 25 m/s, but neither does it increase. The chilling effect is produced by heat being carried away by the wind. The body has to produce the heat to be carried away and a wind of 25 m/s can accomplish this as efficiently as a faster wind. Once the wind removes the heat as fast as it is available at the skin surface, increasing its speed has a negligible effect.

Siple and Passel suggested that the value for L (as expressed in Eq. (2)) with $v = 20$ m/s should be used for all values of v exceeding 20 m/s. It is now generally accepted that wind speeds greater than 20 m/s do not increase the rate of heat loss. We shall accept this upper limit for v as a reasonable one and modify the heat loss formula to include this restriction on v. Then (1) becomes

$$H = \begin{cases} (10.45 + 10\sqrt{v} - v)(33 - T), & 0 \le v \le 20 \\ 35.17(33 - T), & v > 20. \end{cases} \tag{3}$$

Exercises

3. Use the process of completing the square to determine the maximum or minimum values of each of the following functions of x. Also sketch the graph of each function.

   a) $f(x) = -2x^2 + 6x - 7$

   b) $g(x) = x^2 + 4x + 5$

4. Use the process of completing the square to determine the maximum or minimum values of each of the following functions of x. Also sketch the graph of each function.

   a) $f(x) = -x + 6\sqrt{x} - 8$

   b) $g(x) = x - 2\sqrt{x}$

4. Converting to Other Units of Measurement

Public weather reports and forecasts in the United States use degrees Fahrenheit (°F) for reporting temperature and either knots or miles per hour (mi/hr) for reporting wind speed. Thus, it is desirable to convert the heat loss formula into units more familiar to the American public. To do this, we use the conversion equations:

$$°C = (5/9)\,(°F - 32)$$

1 meter ≐ 3.28 feet	or	1 foot ≐ 0.3048 meter
1 m/s ≐ 2.24 mi/hr	or	1 mi/hr ≐ 0.447 m/s

The conversion from mi/hr to m/s is accomplished as follows:

$$1\,\frac{\text{mi}}{\text{hr}} = \frac{5280\text{ ft}}{3600\text{ s}} \times 0.3048\,\frac{\text{m}}{\text{ft}} = 0.447\,\frac{\text{m}}{\text{s}}.$$

Thus, in Eq. (3), v is replaced by $0.447\,v$ and T is replaced by $(5/9)\,(F - 32)$ to give

$$H = (10.45 + 10\sqrt{0.447v} - 0.447\text{v})\,(33 - (5/9)\,(F - 32))$$

$$= \frac{(10.45 + 6.69\sqrt{v} - 0.447\text{v})\,(91.4 - F)}{(9/5)}$$

for the first part of the converted formula. If we change the restriction $0 \le v \le 20$ to $0 \le v \le 45$ and then formulate the second part of Eq. (3), we obtain the converted formula for H:

$$H = \begin{cases} \dfrac{(10.45 + 6.69\sqrt{v} - 0.447\text{v})\,(91.4 - F)}{(9/5)}, & 0 \le \text{v} \le 45 \\ 19.56\,(91.4 - F), & v > 45 \end{cases} \quad (4)$$

where

v is the speed of the wind in mi/hr,
f is the air temperature in °F, and
H is the heat loss in $\text{Kcal/m}^2\text{/hr}$.

Exercises

5. Show that 1 m/s $\doteq$ 1.94 knots. (Hint: 1 knot = 1 nautical mile per hour; 1 nautical mile (n.m.) = 6080 ft.)

6. Convert 25 m/s and 20 m/s to miles per hour. In other words, determine the wind speed in miles per hour that makes the factor $(10.45 + 6.69\sqrt{v} - 0.447v)$ assume its maximum value and express in miles per hour the wind speed that appears as a restriction in Formula (4).

7. Derive the restrictions on v and then the second part of Formula (4).

8. Complete the square of the factor $(10.45 + 6.69\sqrt{v} - 0.447v)$ in Formula (4) to compute the wind speed in miles per hour that makes this factor assume its maximum value.

9. In Section 2, the normal body temperature and the normal skin temperature are given in degrees Celsius. Convert those temperatures to degrees Fahrenheit. Does either temperature appear in Eq. (4)?

5. The Windchill Index

Equations (1), (3), and (4) provide a measure of the heat loss at given temperatures and wind speeds. But the units involved, kilogram calories per square meter per hour, are not familiar to most people. The formulas would be more useful to the general public if the cooling that results from the wind were expressed in familiar units, such as an equivalent temperature felt by the body. This can be accomplished by defining a windchill index (WCI) that bases the cooling on a standard set of conditions.

Empirical evidence suggests that the additional cooling by the wind is minimal for speeds of 4 mi/hr or less. Thus we may use a speed of 4 mi/hr as the standard condition. To define WCI, we then ask what temperature combined with a 4 mi/hr wind produces the same amount of heat loss as a given temperature and wind speed. This temperature is called the Windchill Index or WCI.

More precisely, the WCI for a given temperature F and wind speed v is the temperature that, with a wind speed of 4 mi/hr, produces the same heat loss to exposed flesh as the conditions F and v. To calculate the WCI, we equate the heat loss obtained by using WCI = F and v = 4 in Formula (4) and solve for WCI.

$$\frac{(10.45 + 6.69\sqrt{4} - 0.447(4))\,(91.4 - \text{WCI})}{(9/5)}$$

$$= \frac{(10.45 + 6.69\sqrt{v} - 0.447v)\,(91.4 - F)}{(9/5)}.$$

Thus,

$$22.042\,(91.4 - \text{WCI})$$

$$= (10.45 + 6.69\sqrt{v} - 0.447v)(91.4 - F)$$

and

$$\text{WCI} = 91.4 - \frac{(10.45 + 6.69\sqrt{v} - 0.447v)\,(91.4 - F)}{22.042}.$$

If we round the denominator and add the restrictions from Eq. (4) and the previous paragraphs, we obtain Eq. (5).

$$\text{WCI} = \begin{cases} F, & 0 \le v \le 4 \\ 91.4 - \dfrac{(10.45 + 6.69\sqrt{v} - 0.447v)\,(91.4 - F)}{22}, & \text{for } 4 \le v \le 45 \\ 1.60F - 55, & v > 45. \end{cases} \tag{5}$$

In these equations,

- v is the wind speed in miles per hour,
- F is the air temperature in degrees Fahrenheit, and
- WCI is the equivalent temperature in degrees Fahrenheit felt by exposed skin for the specified temperature and wind speed.

If a wind speed other than 4 mi/hr were chosen as the standard condition, then a different windchill index would result. Although the choice of 4 mi/hr may appear to be arbitrary, evidence suggests that speeds greater than that are necessary for significant cooling to result from wind. Also, a person outside in intense cold is seldom still and the selected speed is representative of what a person moving about would feel.

To compute the Windchill Index for a temperature of 10°F and a wind speed of 20 mi/hr, we substitute these values into Eq. (5) as follows:

$$\text{WCI} = 91.4 - \frac{(10.45 + 6.69\sqrt{20} - 0.447(20))(91.4 - 10)}{22}$$

$$= -25°\text{F}.$$

Consider the significance of this: A person outside in a temperature of 10°F and wind speed of 20 mi/hr would lose as much heat from exposed skin as in a temperature of −25°F with light wind.

Conditions giving rise to a WCI below −20°F are classified as "extreme cold" and are dangerous, as exposed skin will freeze within minutes. The temperature and the duration of the exposure will determine how fast frostbite sets in and how serious it will be. The Windchill Index is a good indicator of the risk of frostbite and is a useful guide to what precautions to take before exposure begins.

Table 1 gives the Windchill Index for selected temperatures and wind speeds. The missing entries are to be supplied in Exercise 10.

Exercises

10. Supply the missing entries in Table 1.

11. Derive the second and third parts of Formula (5).

12. A city reports a Windchill Index of −90°F and a wind speed of 35 mi/hr. Find the air temperature to the nearest degree using Eq. (5).

13. A city reports a Windchill Index of −65°F and an air temperature of −10°F. Find the wind speed to the nearest mi/hr using Eq. (5).

14. Find the Windchill Index that is equivalent to a heat loss of 2300 kilogram calories per square meter per hour. (Refer to Exercise 1).

15. Find the Windchill Index that is equivalent to a heat loss of 2000 kilogram calories per square meter per hour. Refer to Exercise 2.)

16. Discuss the standard condition that uses 4 mi/hr versus 0 mi/hr or some other wind speed. Can you name other "standard conditions" in meteorology or in other branches of science?

17. Derive a formula for the WCI in terms of meters per second and degrees Celsius. Convert 4 mi/hr to m/s and use it as the standard condition.

18. Compute the Windchill Index for a temperature of 5°C and a wind speed of 15 m/sec.

Wind Speed in Miles per Hour

Air Temperature in Degrees Fahrenheit	4	5	10	15	20	25	30	35	40	45	50
35	35	32	22	15	11	8	5	3	2	1	1
30	30	27	16	9	4		-3	-5	-6	-7	-7
25	25	21	9	2	-4	-7	-10	-12	-14	-15	-15
20	20	16	3	-5	-11		-18	-20	-22	-23	-23
15	15	11	-3	-12	-18	-22	-26	-28	-30	-31	-31
10			-9	-18	-25		-33	-36	-38		
5	5	0	-15	-25	-32	-37	-41	-44	-46	-47	-47
0	0	-5	-21	-32	-39	-45	-49	-51	-54	-55	-55
-5	-5	-10	-28	-39	-46	-52	-56	-60	-61	-63	-63
-10	-10	-15	-34	-45	-53	-59	-64	-67	-69	-71	-71
-15	-15	-21	-40	-52	-61		-72	-75	-77	-79	-79
-20	-20	-26	-46	-58	-67	-74	-79	-83	-85	-87	-87
-25	-25	-31	-52	-66	-75	-82	-87	-91	-93	-95	-95
-30	-30	-36	-58	-72	-82	-89	-94	-98	-101	-103	-103
-35	-35	-42	-65	-79	-89	-97	-102	-106	-109	-111	-111
-40	-40	-47	-71	-86	-96		-110	-114	-117	-119	-119
-45	-45	-52	-77	-92	-103	-111	-117	-122	-125	-127	-127
-50	-50	-58	-83	-99	-111	-119	-125	-130	-133	-135	-135

Table 1. Windchill indices for selected temperatures and wind speeds.

6. Bibliography

1. L. J. Battan, *Weather in Your Life*, W. H. Freeman and Company, San Francisco, 1983.

2. A. Court, "Wind Chill," *Bull. Am. Meteor Soc.*, (10)29, 1948, 487-493.

3. D. M. Gates, *Man and His Environment: Climate*, Harper and Row, New York, 1972.

4. H. Landsberg, *Physical Climatology*, 2nd ed., Gray Printing Company, DuBois, Penn., 1958.

5. J. E. Oliver, *Climate and Man's Environment*, Wiley, New York, 1973.

6. J. A. Ruffner and F. E. Bair (eds.), *The Weather Almanac*, 2nd ed., Gale Research Company, Detroit, 1977.

7. B. Schwoegler and M. McClintock, *Weather and Energy*, McGraw-Hill, New York, 1981.

8. P. A. Siple and C. F. Passel, "Measurements of Dry Atmospheric Cooling in Subfreezing Temperatures," *Proc. Am. Philos. Soc.*, (1)89, 1945, 177-199.

9. R. G. Steadman, "Indices of Windchill of Clothed Persons," *J. Appl. Meteor.*, (10), 1971, 674-683.

7. Model Exam

1. Find the maximum value of $f(u) = -4u + 16\sqrt{u} - 6$ and give the value of u for which this maximum value is attained.

2. Express the speed 14 ft/s in the units mi/hr and m/s.

3. A skier is traveling at 30 mi/hr in an air temperature of 10°F. Use Eq. (5) to find the windchill on exposed parts of his or her face.

4. What is the heat loss in ($Kcal/m^2$/hr) under the conditions in Question 3?

5. A heat loss of 1500 $Kcal/m^2/hr$ is occurring at a wind speed of 10 mi/hr. What is the corresponding air temperature in degrees Fahrenheit?

8. Answers to Exercises

1. a) $-54°C$ b) $-32°C$

2. a) $-43°C$ b) $-29°C$

3. a) $f(x) = (-2)(x - 3/2)^2 - 5/2$. Hence f assumes a maximum value of $-5/2$ when $x = 3/2$.

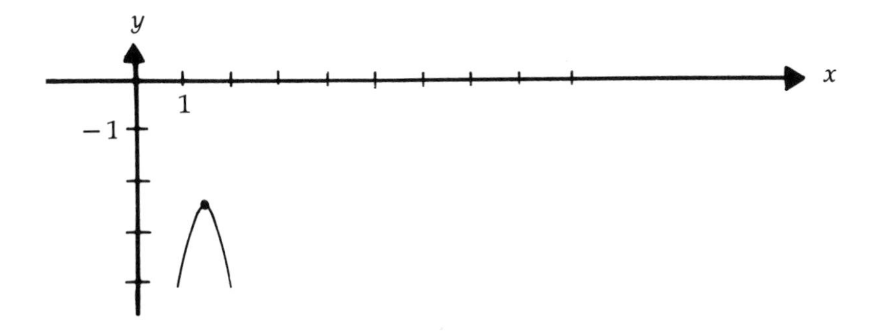

b) $g(x) = (x + 2)^2 + 1$. Thus g assumes a minimum value of 1 when $x = -2$.

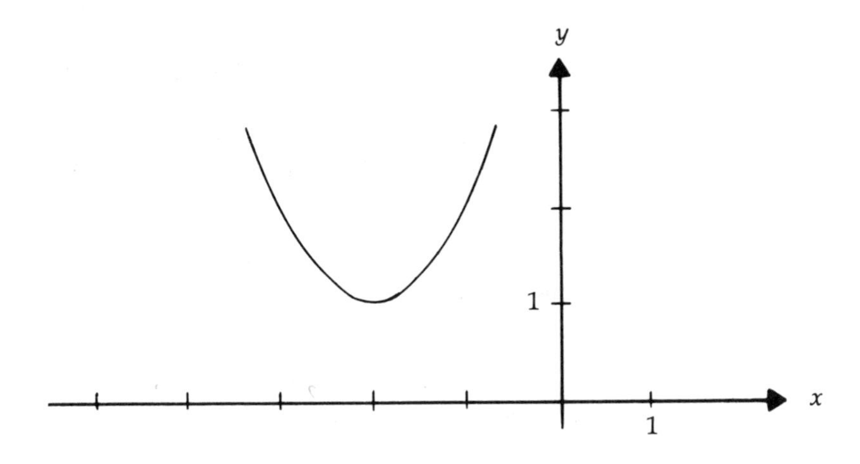

4. a) $f(x) = (-1)(\sqrt{x} - 3)^2 + 1$. So f assumes a maximum of 1 when $x = 9$.

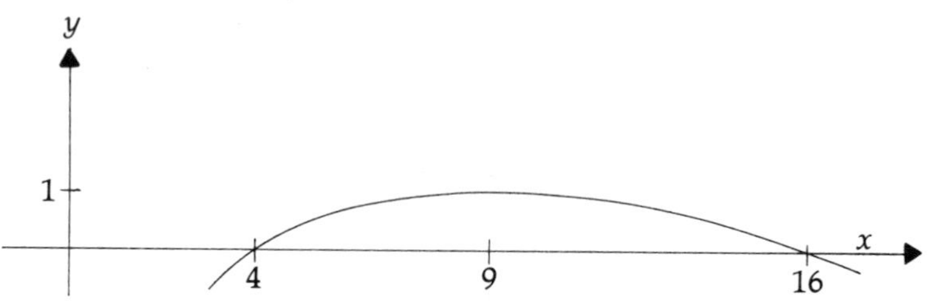

b) $g(x) = (\sqrt{x} - 1)^2 - 1$. So g assumes a minimum value of -1 when $x = 1$.

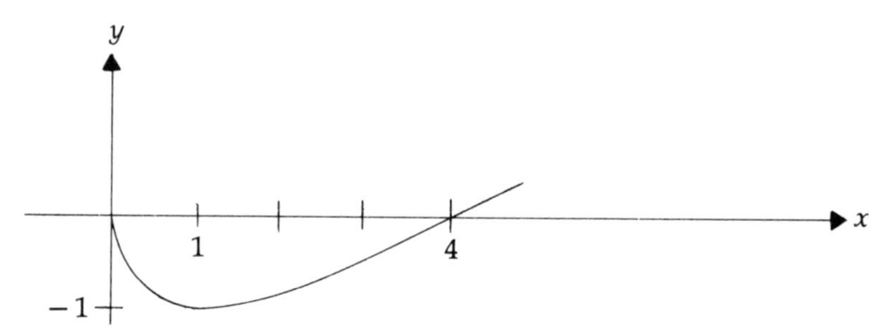

6. a) 25 m/s $\doteq$ 60 mi/hr

b) 20 m/s $\doteq$ 45 mi/hr

7. 20 m/s $\doteq$ 45 mi/hr. If we substitute this value for v in the first part of (4) and simplify, we obtain the second part of (4).

8. $(-0.447)(\sqrt{v} - 7.48)^2 + 35.45$. Thus the maximum value occurs when $\sqrt{v} \doteq 7.48$ or $v \doteq 56$ mi/hr. Note that while 25 m/s $\doteq$ 60 mi/hr, the calculated maximum here occurs when $v \doteq 56$ mi/hr. The difference is due to accumulated round-off errors.

9. 37°C = 98.6°F and 33°C = 91.4°F. Yes, 91.4 appears in the second factor.

			1		
			−15		
10.	10	6	−30	−39	−39
			−67		
			−104		

12

11. To derive the third part, solve

$$\frac{(10.45 + 6.69\sqrt{4} - 0.447(4))\,(91.4 - \text{WCI})}{(9/5)}$$

$$= 19.56(91.4 - F)$$

for WCI.

12. $-25°$F

13. 32 mi/hr

14. $-97°$F

15. $-72°F$

17.

$$\text{WCI} = \begin{cases} T, & 0 \le v \le 1.8 \\ 33 - \dfrac{(10.45 + 10\sqrt{v} - v)}{22}(33 - T), & 1.8 \le v \le 20 \\ 0.63T + 12.3, & v > 20 \end{cases}$$

18. $-11°$C

9. Answers to Model Exam

1. The function f assumes its maximum value of 10 when $u = 4$.

2. 14 ft/s $\doteq$ 9.5 mi/hr $\doteq$ 4.3 m/s

3. -33°F

4. Approximately 1500 Kcal/m^2/hr.

5. Using Eq. (4), we compute the air temperature to be approximately -8°F.

UMAP

Modules in Undergraduate Mathematics and its Applications

Module 659

The Mathematics of Focusing a Camera

Raymond N. Greenwell

Published in cooperation with the Society for Industrial and Applied Mathematics, the Mathematical Association of America, the National Council of Teachers of Mathematics, the American Mathematical Association of Two-Year Colleges, and The Institute of Management Sciences.

INTERMODULAR DESCRIPTION SHEET: UMAP Unit 659

TITLE: THE MATHEMATICS OF FOCUSING A CAMERA

AUTHOR: Raymond N. Greenwell
Department of Mathematics
Hofstra University
Hempstead, NY 11550

CLASSIFICATION: Applications of mathematics to photography.

ABSTRACT: This unit introduces the reader to several photographic concepts, including focal length, f-stop, and depth of field. Using intermediate algebra and trigonometry, we derive the formulas listed in the Kodak Customer Service Pamphlet on the subject. We also show how these formulas can be used in photography.

PREREQUISITES: Intermediate algebra and trigonometry.

The Mathematics of Focusing a Camera

by

Raymond N. Greenwell
Department of Mathematics
Hofstra University
Hempstead, NY 11550

Table of Contents

1. INTRODUCTION 1
2. FOCAL LENGTH 1
3. F-STOPS 2
4. THE FUNDAMENTAL LENS EQUATION 3
5. OTHER FORMS OF THE LENS EQUATION 5
6. DEPTH OF FIELD 8
7. MORE DEPTH OF FIELD 12
8. BIBLIOGRAPHY 16
9. ANSWERS TO EXERCISES 16
10. APPENDIX: THE METRIC SYSTEM FOR LENGTH 20

Modules and Monographs in Undergraduate Mathematics and its Applications Project (UMAP)

The goal of UMAP was to develop, through a community of users and developers, a system of instructional modules in undergraduate mathematics and its applications to be used to supplement existing courses and from which complete courses may eventually be built.

The Project was guided by a National Advisory Board of mathematicians, scientists, and educators. UMAP was funded by a grant from the National Science Foundation and is now supported by the Consortium for Mathematics and Its Applications, Inc. (COMAP), a nonprofit corporation engaged in research and development in mathematics education.

This material was prepared with the partial support of National Science Foundation Grant No. SED80-07731. Recommendations expressed are those of the authors and do not necessarily reflect the views of the NSF or the copyright holder.

1. Introduction

Before the spring of 1982, my only camera was an Instamatic, a simple device that I recklessly dragged through the mud on my backpacking trips. Then I bought my first 35mm camera. I was amazed at the number of buttons and dials. Once I got the box open, the camera inside seemed even more complex. I soon learned to take pictures I could never have taken with an Instamatic, such as an inspiring close-up of the inside of the lens cover.

Another item I obtained at the same time was the Kodak Customer Service Pamphlet AA-26, entitled *Optical Formulas and Their Applications.* Here was something of interest to a mathematician. Included was not only the fundamental lens equation, sometimes called the thin lens equation, but also others involving focal length, magnification, and depth of field. In this module we shall derive those equations and present some of their applications.

2. Focal Length

The first term we need is the *focal length* of the camera lens, which we will denote F. If parallel rays of light approach the lens, they are focused at a point, as shown in Fig. 1. The distance from this point to the lens is F. If light rays from a point infinitely far away approach the lens, they would be parallel, as in Fig. 1. For all practical purposes, infinity isn't very far away. On a standard camera lens (one with a focal length of about 50 mm) we can consider anything more than 20 m away to be at infinity. (If you are unfamiliar with the metric system, see Section 10.)

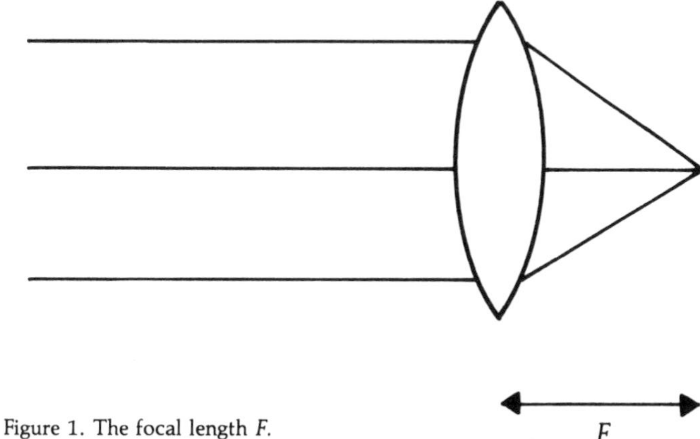

Figure 1. The focal length F.

The easiest way to find the focal length of a lens is to read it on the front or side of the lens. The lens that came with my camera has a focal length of 50 mm. The wide angle lens I bought has $F = 28$ mm, while the zoom lens has an F that varies from 80 mm to 200 mm. We shall discuss later the effect of the focal length on the pictures you take.

3. F-Stops

The *f-stop*, or *f-number*, of a lens is the focal length divided by the diameter of the lens opening, which we will denote by L. Mathematically, this says

$$f = \frac{F}{L}. \tag{1}$$

The f-stop is written as **f** followed by a slash followed by the numerical value, such as **f/8**. For a given focal length F, the f-stop is made smaller by making the lens opening, known as the *aperture*, larger. This increases the amount of light that falls on the film, which is what we want under low light conditions.

Unfortunately, as we shall see later, increasing the aperture also tends to put more of the picture out of focus. This might be desirable if we want to focus attention on one part of the picture and put everything else out of focus. Otherwise, we can let in more light by using a slower shutter speed, which causes the aperture to be open for a longer period of time. The disadvantage of this is that any movement of the camera or subject results in a blurry picture. Another solution is to use film with a higher ASA (or ISO) number, which means it requires less light. The drawback is that the pictures will then have a grainier texture. In photography, as in life, there are no free lunches.

Most lenses have the following f-stops: **f/2**, **f/2.8**, **f/4**, **f/5.6**, **f/8**, **f/11**, **f/16**, and **f/22**. These are the full stops, but there are also half stops in between these. The full stops are set so that each lets in half as much light as the previous stop. The amount of light let in is proportional to the area of the aperture, which is roughly circular, and the area of a circle is proportional to the diameter squared. ($A = \pi r^2 = \pi L^2/4$, where $L = 2r$.) Thus, the diameter must be reduced by $1/\sqrt{2}$ to reduce the area by

1/2. According to Eq. (1), if L is multiplied by $1/\sqrt{2}$, then f is multiplied by $\sqrt{2}$. For example, if the first f-stop is 2, the next one is $2\sqrt{2} = 2.828$, which is rounded to 2.8. We can continue this process and compute the exact values of all the full stops.

Exercise

1. Find the exact value of all the full f-stops. Which two of the values given previously are rounded incorrectly?

4. The Fundamental Lens Equation

We shall now derive the fundamental lens equation. Let u be the distance from the object focused upon to the lens, and let v be the distance from the lens to the film. The light rays travel through the paths indicated in Fig. 2. We denote the height of the object by h and the height of the image on the film by h'. If u approaches infinity, F and v become equal. (Recall the definition of F in Section 2.) Otherwise, the lens must be farther from the film than F, and this extra distance is denoted x. This is why the lens housing becomes longer when you focus on a close object.

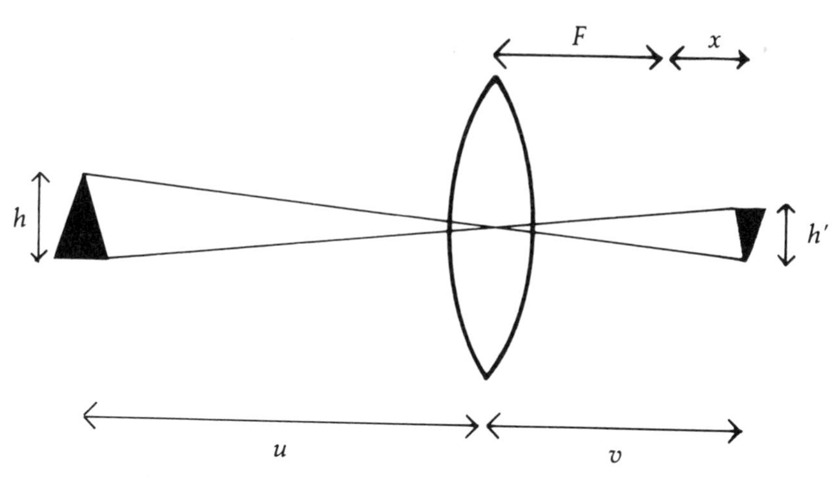

Figure 2. Quantities involved in the fundamental lens equation.

Now let us focus our attention at a point on the top of the object denoted P (see Fig. 3). The light rays from this point do not all travel straight to the point at the bottom of the image, denoted P'. They travel out in all directions, and all rays from P that hit the lens are focused on P'. Two such rays are shown in Fig. 3, one traveling horizontally to point A, and the other traveling to a point level with P', which we call B. Because parallel rays of light are focused at a distance F from the lens, the points C and C' must both be a distance F from the lens. If we denote the point in the middle of the lens by D, then $\overline{CD} = \overline{C'D} = F$.

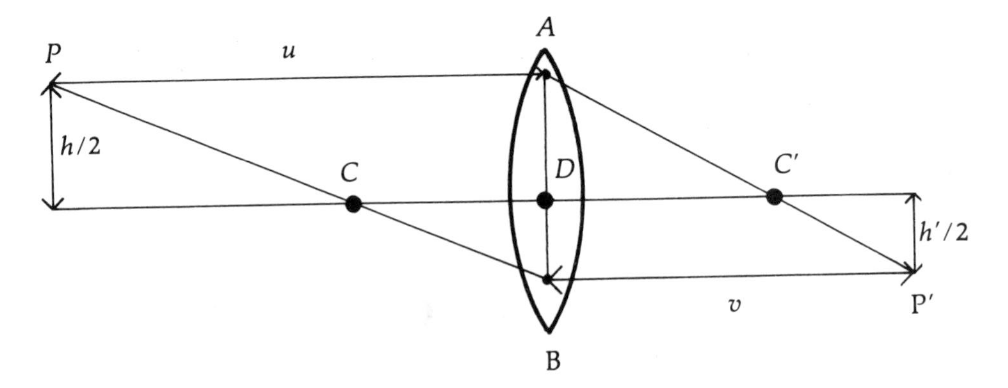

Figure 3. Rays of light focused by the lens.

Since triangles PAB and CDB are similar, we have

$$\frac{\overline{AB}}{\overline{PA}} = \frac{\overline{DB}}{\overline{CD}}. \tag{2}$$

Recalling that $\overline{PA} = u$, and seeing from Fig. 3 that

$$\overline{AB} = h/2 + h'/2,$$

Eq. (2) is transformed into

$$\frac{h/2 + h'/2}{u} = \frac{h'/2}{F}. \tag{3}$$

Next, we note that triangle ABP' and ADC' are similar, so we have

$$\frac{\overline{AB}}{\overline{PB}} = \frac{\overline{AD}}{\overline{C'D}}, \tag{4}$$

which becomes

$$\frac{h/2 + h'/2}{v} = \frac{h/2}{F}. \tag{5}$$

Adding Eqs. (3) and (5) gives

$$\frac{h/2 + h'/2}{u} = \frac{h/2 + h'/2}{v} = \frac{h'/2}{F} + \frac{h/2}{F}. \tag{6}$$

Dividing by $h/2 + h'/2$ yields

$$\frac{1}{u} + \frac{1}{v} = \frac{1}{F}. \tag{7}$$

This is the important formula we have been seeking.

Exercises

2. If a 50 mm lens is focused on an object 5 m away, how far is the lens from the film?

3. We earlier stated that if $u \geq 20$ m for a 50 mm lens, the object is considered to be at infinity, so v and F are essentially equal. Find v for a 50 mm lens if $u = 20$ m.

5. Other Forms of The Lens Equation

The Kodak booklet has several formulas immediately after the lens equation that are described as "more directly useful." They first define the magnification m:

$$m = \frac{h'}{h}. \tag{8}$$

If the image on the film is equal in size to the object being photographed, then $m = 1$. In most cases the image on the film is much smaller, so $m < 1$.

From the similarity of triangles in Fig. 2, we see that we can also express the magnification as

$$m = \frac{v}{u}. \tag{9}$$

Exercises

4. Derive the equations

$$u = \frac{Fv}{v - F} \tag{10}$$

and

$$v = \frac{Fu}{u - F}. \tag{11}$$

5. Use Eq. (10) to eliminate u from Eq. (9) and get

$$m = \frac{v - F}{F}. \tag{12}$$

6. Use Eq. (11) to eliminate v from Eq. (9) and get

$$m = \frac{F}{u - F}. \tag{13}$$

Notice from Eq. (13) that m gets smaller as u gets larger. If an object really could be infinitely far away, then the magnification would be zero. For example, if you photograph a golf ball from very far away, the image on the film virtually disappears.

We can draw an important conclusion from Eq. (13). If F is made larger, the numerator of the right-hand side increases while the denominator decreases, resulting in a larger value of m. In other words, a lens with a large focal length has a large magnification. If I want to photograph a moose that is far away, I use my telephoto lens with $F = 200$ mm so that the animal will have a large image in the picture. On the other hand, if I want to photograph the Grand Canyon, I use my wide angle lens with $F = 28$ mm, and the small magnification allows more of the Grand Canyon to fit my picture.

Now let's derive another useful equation. If we solve Eq. (9) for v, we get $v = mu$. Putting this into Eq. (7) gives

$$\frac{1}{mu} + \frac{1}{u} = \frac{1}{F}$$

or

$$\frac{1}{u}\left(\frac{1}{m} + 1\right) = \frac{1}{F}.$$

Multiplying by uF yields

$$u = F\left(\frac{1}{m} + 1\right). \tag{14}$$

As an example of how we can use Eq. (14), suppose we want to photograph a 180 cm person with a 50 mm lens so that the image completely fills up the picture the long way. Since 35 mm film has an image 24 mm by 36 mm, the magnification is 36 mm/180 cm = 0.02. Putting this and F = 50 mm = 0.05 m into Eq. (13) yields u = 2.55 m. Thus our subject must stand 2.55 m away.

Exercises

7. Find out how far away our subject must stand if we want him to completely fill up the image the short way and we use our wide angle lens, with a focal length of 28 mm.

8. What focal length should we use to photograph the Empire State Building, which is 381 m high, if we are standing 200 m away? Assume that we want the building to almost fill up the picture the long way, with 2 mm of extra space at the top and bottom of the negative.

9. The Kodak booklet also lists the following formulas. Derive each of them:

   a) $v = (m + 1)F$

   b) $u = \left(\frac{1}{m} + 1\right)F$

   c) $u + v = \frac{(m + 1)^2}{m} F$

We can also compute x, the distance the lens must move from the infinity setting. From Fig. 1, we see that

$$x = v - F,$$

and using Eq. (11) gives

$$x = \frac{Fu}{u - F} - F\,\frac{(u - F)}{(u - F)} = \frac{F^2}{u - F}. \tag{15}$$

10. How much must a 50 mm lens move from the infinity position to focus on an object 1.5 m away?

6. Depth of Field

When the camera is focused at a given distance, points not too much nearer or farther away are still roughly in focus. The difference between the nearest and farthest points from the camera that are still in focus is called the *depth of field.* The exact location of those two points, called the near and far limits of the depth of field, is a matter of opinion, since objects do not suddenly become completely out of focus when they reach the limits of the depth of field. Furthermore, what is considered in focus for a 3 x 5 inch print may not be acceptable for an 8 x 12 inch enlargement.

General standards for the depth of field depend upon something called the circle of confusion. This is not one of those perplexing traffic circles in Washington, D.C. The circle of confusion is the circular image on the film of a point not exactly in focus. If the circle is small enough, the point appears to be in focus, and is therefore in the depth of field. Fig. 4 shows why a point can result in a circle, rather than a point, on the film. Point P is exactly in focus, resulting in a point P' on the film. Point Q is closer than point P, and the lens focuses it at a point Q' behind the film. The film receives a circle of diameter d; it looks like a line segment in Fig. 4 because we are viewing it from the side. A similar event occurs if Q is further from the lens than P, as shown in Fig. 5. According to the Kodak booklet, the most widely used value of d for 35 mm film is 0.002. (Some books recommend a smaller value.) We will derive formulas for the near and far limits of the depth of field as a function of d.

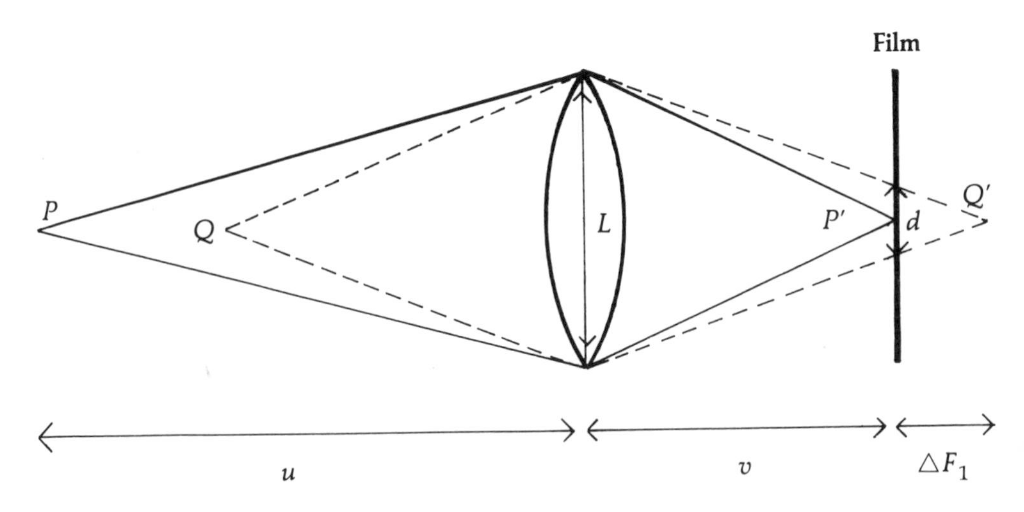

Figure 4. Origin of the circle of confusion.

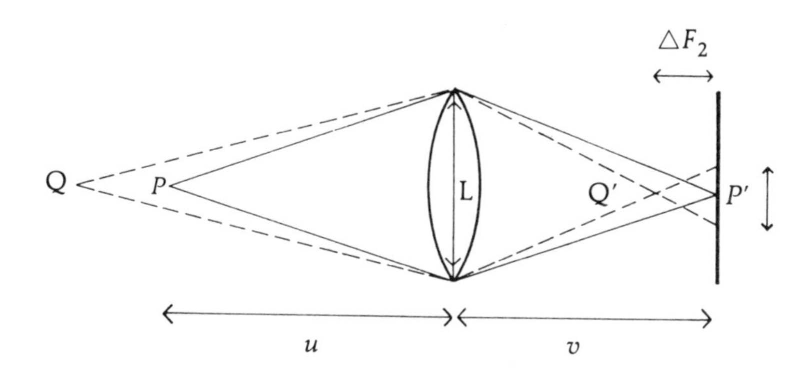

Figure 5. The circle of confusion for a point farther away.

Let $\triangle F_1$ be the distance from P' to Q', as shown in Fig. 4. Then by similarity of the triangle with Q' as a vertex and L as a base to the triangle with Q' as vertex and d as a base, we have

$$\frac{\triangle F_1}{d} = \frac{v + \triangle F_1}{L}. \tag{16}$$

The quantity $\triangle F_1$ is very small compared with v, so we will drop it. Recall from Fig. 2 that $v = F + x$, so Equation (16) becomes

$$\frac{\triangle F_1}{d} = \frac{F + x}{L}. \tag{17}$$

We now use similarity of the two triangles in Fig. 5 with Q' as a vertex, one with L as a base and the other with d as a base, to get

$$\frac{\triangle F_2}{d} = \frac{v - \triangle F_2}{L}. \tag{18}$$

and by the same reasoning as before, this becomes

$$\frac{\triangle F_2}{d} = \frac{F + x}{L}. \tag{19}$$

Comparing Eqs. (17) and (19), we see that $\triangle F_1$ and $\triangle F_2$ are equal, at least to the degree of our approximation, so we will denote them both by $\triangle F$. Either equation tells us

$$\triangle F = \frac{d}{L}(F + x). \tag{20}$$

Let u_1 be the near limit of the depth of field, corresponding to the distance of point Q from the lens in Fig. 4. Let u_2 be the far limit of the depth of field, corresponding to the distance of point Q from the lens in Fig. 5. Equation (10) gives us a formula to compute these two quantities. To get u_1, we substitute $v + \triangle F$ for v in Eq. (10), and then let $v = F + x$, yielding

$$\begin{aligned} u_1 &= \frac{F(v + \triangle F)}{(v + \triangle F) - F} \\ &= \frac{F(F + x + \triangle F)}{F + x + \triangle F - F} \\ &= \frac{F(F + x + \triangle F)}{x + \triangle F}. \end{aligned} \tag{21}$$

As before, the $\triangle F$ in the numerator is small compared with $F + x$, so we drop it. We do not drop it in the denominator, since x is not very big. Thus Eq. (21) becomes

$$u_1 = \frac{F(F + x)}{x + \triangle F}. \tag{22}$$

Exercise

11. Use reasoning similar to the reasoning given above to derive the equation

$$u_2 = \frac{F(F + x)}{x - \triangle F}. \tag{23}$$

We would like to get rid of x and $\triangle F$, whose values we usually do not know, and write Eqs. (22) and (23) in terms of more familiar quantities. To do this, we now introduce one more new concept, the *hyperfocal distance.* Denoted by H, the hyperfocal distance is just the near limit of the depth of field when the lens is focused at infinity. We saw earlier that when u becomes infinite, x becomes 0. Then Eq. (22) tells us that the near limit of the depth of field is

$$H = \frac{F^2}{\triangle F}. \tag{24}$$

We can eliminate $\triangle F$ by using Eq. (20), with $x = 0$, giving

$$H = \frac{F^2}{(d/L)F} = \frac{FL}{d}. \tag{25}$$

A better equation for H does not use L, which we have to compute, and instead uses f, which we can read off the side of the lens. We can use Eq. (11) to change from L to f, yielding

$$H = \frac{F(F/f)}{d} = \frac{F2}{fd}. \tag{26}$$

We will now eliminate $\triangle F$ from Eq. (22) by using Eq. (24), which tells us that

$$\triangle F = \frac{F2}{H}. \tag{27}$$

Putting this into Eq. (22) gives

$$\begin{aligned} u_1 &= \frac{F(F + x)}{x + F2/H} \\ &= \frac{HF(F + x)}{Hx + F2}. \end{aligned} \tag{28}$$

Finally, we need an equation to eliminate x. Looking back over the multitude of equations we have derived, we see that Eq. (15) will do the job. The result is

$$\begin{aligned} u_1 &= \frac{HF\ (F + \dfrac{F2}{u - F})}{H(\dfrac{F2}{u - F}) + F2} \times \frac{u - F}{u - F} \\ &= \frac{HF\ (F\,(u - F) + F2)}{HF2 + F2(u - F)} \\ &= \frac{HF2(u - F + F)}{F2(H + u - F)} \end{aligned}$$

$$= \frac{Hu}{H + (u - F)}. \tag{29}$$

This is the equation for the near limit of the depth of field as given in the Kodak booklet.

Exercises

12. Following the procedure above, derive the equation for the far limit of the depth of field:

$$u_2 = \frac{Hu}{H - (u - F)}. \tag{30}$$

13. Letting $d = 0.002$ inches, find the hyperfocal distance of a 50 mm lens set at **f/8**.

14. Use your answer from Exercise 13 to find the near and far limits of the depth of field for a 50 mm lens set at **f/8** focused on a point 5 m away.

15. If I set my 50 mm lens at **f/8** and focus on 2 m, little numbers on the side indicate that the near and far limits of the depth of field are roughly 1.7 m and 2.6 m, respectively. What are the makers of my lens using for the diameter of the circle of confusion?

7. More Depth of Field

The Kodak booklet lists two additional formulas for the depth of field. They involve the angular size of the circle of confusion, denoted by θ in Fig. 6. The Kodak booklet suggests letting θ be 2 minutes of arc, or $1/30°$.

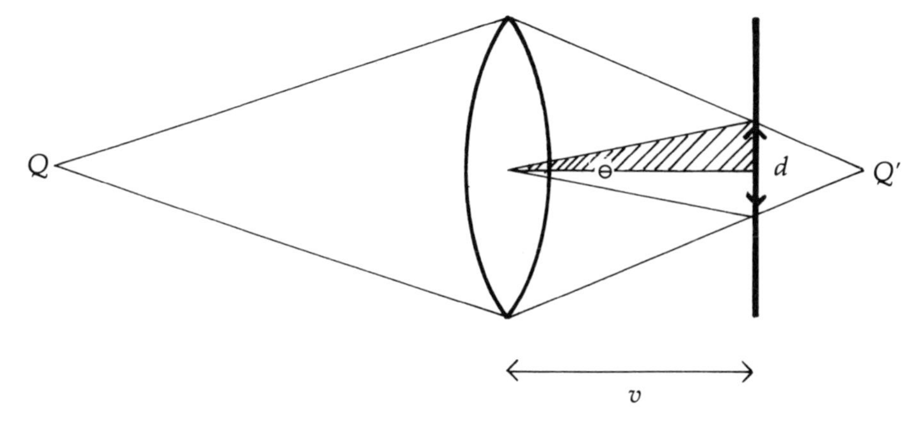

Figure 6. Angular size of the circle of confusion.

We can see from the shaded triangle that

$$\tan \frac{\Theta}{2} = \frac{(d/2)}{v}. \tag{31}$$

It would be easier to write Eq. (31) in terms of the tan Θ. Using the double angle formula for tangents, we can write

$$\tan \Theta = \frac{2 \tan \frac{\Theta}{2}}{1 - \tan^2 \frac{\Theta}{2}}. \tag{32}$$

But $\tan^2(\Theta/2)$ is very small, compared with 1, for such a small Θ, so Eq. (32) tells us that the equation

$$\tan \frac{\Theta}{2} = \frac{1}{2} \tan \Theta \tag{33}$$

is approximately true for small Θ. Then Eq. (31) can be written

$$\tan \Theta = \frac{d}{v}. \tag{34}$$

Using Eq. (11) to eliminate v from Equation (34),

$$\tan \Theta = \frac{d(u - F)}{Fu}$$

or

$$u - F = \frac{Fu(\tan \theta)}{d}. \tag{35}$$

We will need this in a few moments. First, we will note that the new formulas in the Kodak booklet give the near and far limits of the depth of field measured from the plane focused upon, rather than from the camera. (See Fig. 7.) Letting w_1 and w_2 be these quantities, we have

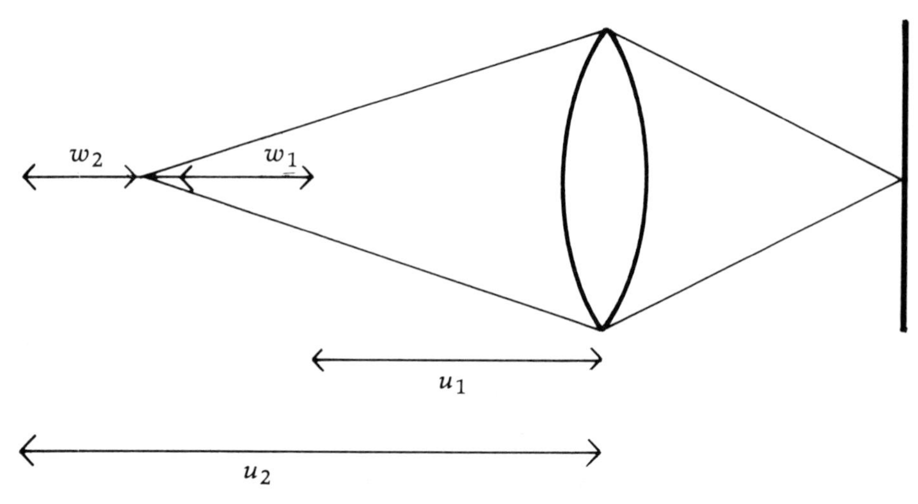

Figure 7. Measuring depth of field from plane focused upon.

$$\begin{aligned} w_1 &= u - u_1 \\ &= u - \frac{Hu}{H + u - F} \\ &= \frac{u(H + u - F) - Hu}{H + u - F} \\ &= \frac{u(u - F)}{H + (u - F)}. \end{aligned} \tag{36}$$

Exercise

16. Derive the equation for the far limit of the depth of field as measure from the plane focused upon:

$$w_2 = \frac{u(u - F)}{H - (u - F)}. \tag{37}$$

Now we can use Eq. (35) to eliminate $u - F$ from Eq. (36), yielding

$$w_1 = \frac{u \dfrac{Fu(\tan \theta)}{d}}{H + \dfrac{Fu(\tan \theta)}{d}}$$

$$= \frac{Fu^2(\tan \theta)}{Hd + Fu(\tan \theta)}. \tag{38}$$

Finally, we get rid of H by using Eq. (25) to write $Hd = FL$. Then Eq. (37) becomes

$$w_1 = \frac{Fu^2(\tan \theta)}{FL + Fu(\tan \theta)}$$

$$= \frac{u^2(\tan \theta)}{L + u(\tan \theta)}. \tag{39}$$

This is the formula in the Kodak booklet.

Exercises

17. Following the procedure above, derive the form for the far limit of the depth of field as measured from the plane focused upon:

$$w_2 = \frac{u^2(\tan \theta)}{L - u(\tan \theta)}. \tag{40}$$

18. Using $\theta = 1/30°$, find w_1 and w_2 for 50 mm lens set at **f/8** focused on a point 5 m away. How does this compare with your answer to Exercise 14?

8. Bibliography

The Kodak Customer Service Pamphlet AA-26, *Optical Formulas and Their Applications*, costs 10 cents and can be purchased at most photography stores or by writing to Eastman Kodak Company, Consumer Markets Division, 343 State Street, Rochester, New York 14650. The following books may also be of interest:

Adams, Ansel, *The Camera*, New York Graphic Society, 1980.

Cox, A.;, *Optics, the Techniques of Definition*, Focal Press, 1966.

Magill, Arthur A., "Still Cameras," in *Applied Optics and Optical Engineering Vol. IV, Optical Instruments*, edited by Rudolf Kingslake, Academic Press, 1967.

Neblette, C. B. and Murray, Allen E., *Photographic Lenses*, Morgan and Morgan, Inc., 1973.

9. Answers to Exercises

1. $2 \times (\sqrt{2})^2 = 4$, $2 \times (\sqrt{2})^3 = 5.66$, $2 \times (\sqrt{2})^4 = 8$, $2 \times (\sqrt{2})^5 = 11.31$, $2 \times (\sqrt{2})^6 = 16$, $2 \times (\sqrt{2})^7 = 22.62$. The third f-stop should be rounded to 5.7 rather than 5.6, and the last one should be rounded to 23 rather than 22.

2. 5 m = 5000 mm.
   $1/5000 + 1/v = 1/50 \Rightarrow 1/v = 1/50 - 1/5000 = 99/5000$,
   $v = 5000/99 = 50.505$ mm.

3. 20 mm = 20000 mm.
   $1/20000 + 1/v = 1/50 \quad 1/v = 1/50 - 1/20000 = 399/20000$, $v = 20000/399 = 50.125$ mm.

4. $1/u = 1/F - 1/v = (v - F)/vF$, so $u = vF/(v - F)$. Eq. (11) is derived similarly.

5. $$m = \frac{v}{u} = \frac{v}{\left(\frac{vf}{v - F}\right)} = \frac{v(v - F)}{vF} = \frac{v - F}{F}.$$

6. $m = \frac{v}{u} = \frac{\left(\frac{uF}{u - F}\right)}{u} = \frac{uF}{u(u - F)} = \frac{F}{u - F}.$

7. $m = 2.4 \text{ cm}/180 \text{ cm} = 0.01333.$

$u = 0.028\left(\frac{1}{0.01333} + 1\right) = 2.13 \text{ m}.$

8. $h' = 36 - 2 \times 2 = 32 \text{ mm} = 0.032 \text{ m}.$

$m = 0.032/381 = 8.399 \times 10^{-5}$

$= F/(200 - F) \quad 0.016798 - 8.399 \times 10^{-5}F = F,$

$0.016798 = F(1 + 8.399 \times 10^{-5}),$

$F = 0.017698/(1 + 8.399 \times 10^{-5}) = 0.017 \text{ m} = 17 \text{ mm}.$

9. **a)** From Eq. (12), $mF = v - F$, so $v = mF + F = (m + 1)F.$

b) From Eq. (13),
$mu - mF$, so $mu = F + mF$, $u = (F/m) + F = (1/m + 1)F.$

c) Adding the answers to (**a**) and (**b**),

$u + v = (m + 1)F + \left(\frac{1}{m} + 1\right)F$

$= \left(m + 2 + \frac{1}{m}\right)F = \left(\frac{m^2 + 2m + 1}{m}\right)F = \frac{(m + 1)^2}{m}F.$

10. $u = 1.5 \text{ m} = 1500 \text{ mm}.\ x = \frac{50^2}{1500 - 50} = 1.7 \text{ mm}.$

11. $u_2 = \frac{F(v - \triangle F)}{(v - \triangle F)} = \frac{F(F + x - \triangle F)}{F + x - \triangle F - F}$

$= \frac{F(F + x - \triangle F)}{x - \triangle F} \cong \frac{F(F + x)}{x - \triangle F}$

12. $u_2 = \dfrac{F(F + x)}{x - \dfrac{F^2}{H}} = \dfrac{HF(F + x)}{Hx - F^2} = \dfrac{HF\left(F + \dfrac{F^2}{u - F}\right)}{H\left(\dfrac{F^2}{u - F}\right) - F^2}$

$$= \frac{HF(F(u - F) + F^2)}{HF^2 - F^2(u - F)} = \frac{HF^2(u - F + F)}{F^2(H - (u - F))} = \frac{Hu}{H - (u - F)}.$$

13. $d = 0.002 \text{ in} \times 25.4 \text{ mm/in} = 0.0508 \text{ mm}.$

$$H = \frac{50^2}{8 \times 0.0508} = 6152 \text{ mm} = 6.2 \text{ m}.$$

14. $u_1 = \dfrac{6.2 \times 5}{6.2 + 5 - 0.05} = 2.8 \text{ m}.$

$$u_2 = \frac{6.2 \times 5}{6.2 - (5 - 0.05)} = 25 \text{ m}.$$

15. Using Eq. (29), $1.7 = \dfrac{H \times 2}{H + 2 - 0.05}$ $\quad 1.7H + 3.315 = 2H,$

$0.3H = 3.315,\ H = 3.315/0.3 = 11 \text{ m}.$

Checking this with Eq. (30),

$2.6 = \dfrac{H \times 2}{H - (2 - 0.05)}$ $\quad 2.6H - 5.07 = 2H,\ 0.6H = 5.07,$

$H = 5.07/0.6 = 8.5$ m. Using the first value in Eq. (26),
$d = F^2/fH = 50^2/8 \times 11000 = 0.028/25.4 \text{ mm/in} = 0.001 \text{ in}.$

Using the second value,
$d = 50^2/8 \times 8500 = 0.037 \text{ mm}/25.4 \text{ mm/in.} = 0.001 \text{ in}.$

16. $w_2 = u_2 - u = \dfrac{Hu}{H - (u - F)} - u = \dfrac{Hu - u(H - u + F)}{H - (u - F)}$

$$= \frac{u(u - F)}{H - (u - F)}.$$

17. $$w_2 = \frac{u\dfrac{Fu(\tan\theta)}{d}}{H - \dfrac{Fu(\tan\theta)}{d}} = \frac{Fu^2(\tan\theta)}{Hd - Fu(\tan\theta)} = \frac{Fu^2(\tan\theta)}{FL - Fu(\tan\theta)}$$

$$= \frac{u^2(\tan\theta)}{L - u(\tan\theta)}.$$

18. $$L = \frac{0.05\ m}{8} = 0.00625\ \text{m}.$$

$$w_1 = \frac{5^2 \tan\dfrac{1^\circ}{30}}{0.00625 + 5 \times \tan\dfrac{1^\circ}{30}} = 1.6\ \text{m}.$$

$$w_2 = \frac{5^2 \tan\dfrac{1^\circ}{30}}{0.00625 - 5 \times \tan\dfrac{1^\circ}{30}} = 4.4\ \text{m}.$$

This corresponds to $u_1 = u - w_1 = 5 - 1.6\ \text{m} = 3.4\ \text{m}$ and $u_2 = u + w_2 = 5 + 4.4\ \text{m} = 9.9\ \text{m}$. This is much narrower than the depth of field in Exercise 14.

10. Appendix: The Metric System for Length

The standard unit of length in the metric system is the meter, which is 39.37 inches, or slightly more than a yard. Conversely, one inch is 1/39.37 meters, or 0.0254 meters. Therefore, to convert inches to meters, we multiply the quantity in inches by 0.0254 m/in. For example, 42 in = 42 × 0.0254 = 1.07 m (rounded off). To convert meters to inches, divide by 0.0254 m/in. For example, 2.1 m = 2.1/0.0254 = 83 in.

A meter is divided into 100 centimeters, each of which is divided into 10 millimeters. There are 1000 millimeters in a meter. It is easy to convert back and forth from millimeters to meters, as is frequently done in this unit. Just move the decimal point three places to the right to go from meters to millimeters, and three places to the left to go from millimeters to meters. For example, 50 mm = 0.050 m, and 1.6 m = 1600 mm.

Since there are 0.0254 meters in an inch, there are 25.4 mm in an inch. We can thus convert inches to millimeters by multiplying by 25.4 mm/in. For example, 12 in. = 12 × 25.4 = 305 mm, 50 mm = 50/25.4 = 1.97 in.

UMAP

Modules in Undergraduate Mathematics and its Applications

Module 661

Where are the Russian and Chinese Missiles Coming From?

Sidney H. Kung

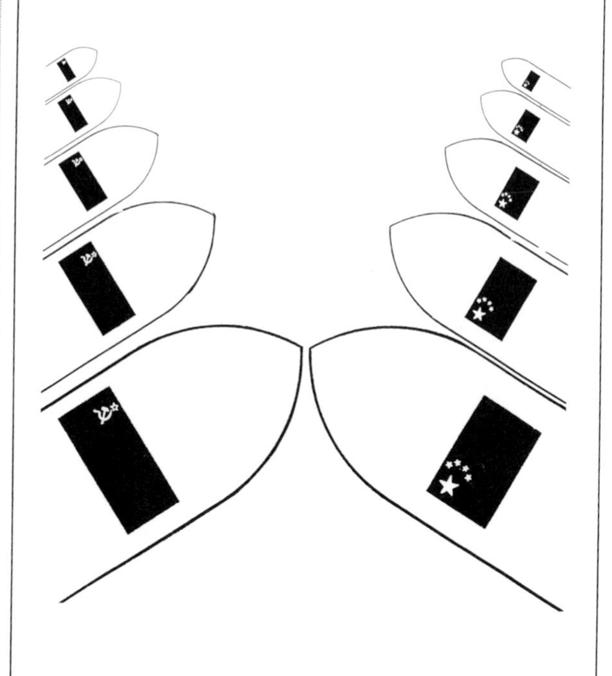

Published in cooperation with the Society for Industrial and Applied Mathematics, the Mathematical Association of America, the National Council of Teachers of Mathematics, the American Mathematical Association of Two-Year Colleges, and The Institute of Management Sciences.

INTERMODULAR DESCRIPTION SHEET:	UNIT 661
TITLE:	WHERE ARE THE RUSSIAN AND CHINESE MISSILES COMING FROM?
AUTHOR:	Sidney H. Kung
MATH FIELD:	Calculus
APPLICATION FIELD:	Military Science
TARGET AUDIENCE:	Students in a multivariable calculus course
ABSTRACT:	New warheads and long-range rockets have been frequently tested by the USSR, the USA, and, in recent years, by the PRC. To a serious applied mathematics student, there is a different kind of message to be explored and interpreted. Why is there an impact zone? How reliable were reported figures? How accurate were the missiles in reaching their targets? And, at what locations were the missiles launched? In this module, we use physics and a basic knowledge of vector algebra to obtain some reasonable answers.
PREREQUISITES:	Dot products, cross products.

Where are the Russian and Chinese Missiles Coming From?

Sidney H. Kung
Department of Mathematics
Jacksonville University
Jacksonville, FL 32211

Table of Contents

1. INTRODUCTION 1
2. THE RUSSIAN MISSILE 2
   2.1 Determining the Test Site 2
   2.2 The Coordinates of Points of Impact 2
   2.3 A Scaler Triple Production Application 4
   2.4 The Range 6
   2.5 The Direction 7
   2.6 The Aiming Accuracy 8
3. THE CHINESE MISSILE 9
4. ANSWERS TO EXERCISES 10

"Dedicated to J.E.S., a special friend"

MODULES AND MONOGRAPHS IN UNDERGRADUATE MATHEMATICS AND ITS APPLICATIONS PROJECT (UMAP)

The goal of UMAP was to develop, through a community of users and developers, a system of instructional modules in undergraduate mathematics and its applications to be used to supplement existing courses and from which complete courses may eventually be built.

The Project was guided by a National Advisory Board of mathematicians, scientists, and educators. UMAP was funded by a grant from the National Science Foundation and is now supported by the Consortium for Mathematics and Its Applications, Inc. (COMAP), a nonprofit corporation engaged in research and development in mathematics education.

We would like to thank Professor Carroll O. Wilde and Professor Roger H. Moritz for reviewing the manuscript and making many valuable suggestions.

This material was prepared with the partial support of National Science Foundation Grant No. SPE8304192. Recommendations expressed are those of the author and do not necessarily reflect the views of the NSF or the copyright holder.

1. Introduction

You may have read the following news items:

In the May 12, 1980 issue of the *Aviation Week and Space Technology*: "Chinese CSS-X-4 8000 mile-range ICBM will be given maximum-range tests with impact near Kiribati, Tuvalu, in the Solomon Islands region of the Pacific, the Chinese government has told the Australian and New Zealand governments."

In the May 19, 1980 *The New York Times*: "Chinese News Agency said that China had successfully launched its first ICBM into the target area with a diameter of 70 nautical miles centered on 7 degrees South Latitude, and 171 degrees 33 minutes East Longitude. The missile is believed to have been launched from a test site in northern China, traveling roughly 6000 miles before landing."

In the September 1, 1961 *The Chinese People's Daily*: "The Soviets have announced that a powerful missile test is underway. The last-stage rocket will enter the area, approximately a rectangle, which is determined by the following geographical coordinates:

Table 1

	P_1	P_2	P_3	P_4
North Latitude	10°20′	11°30′	9°10′	8° 5′
West Longitude	170°30′	167°55′	166°45′	169°20′ "

New warheads and long-range rockets have been frequently tested by the USSR, the USA, and, in recent years, by the PRC. No doubt, such news must have political, strategic, and technological implications. To a serious applied mathematics student, there is a different kind of message to be explored and interpreted. Why is there an impact zone (two types)? How reliable were the reported figures (two different sets) in the range of the Chinese missile? How accurate were the missiles in reaching their targets? And, even more importantly, at what locations were the missiles launched? It may seem impossible to answer these questions readily because so little information is given. However, in this module, we shall analyze the "old" news mathematically by using physics and a basic knowledge of vector algebra to obtain some reasonable answers.

2. The Russian Missile

2.1 Determining the Test Site

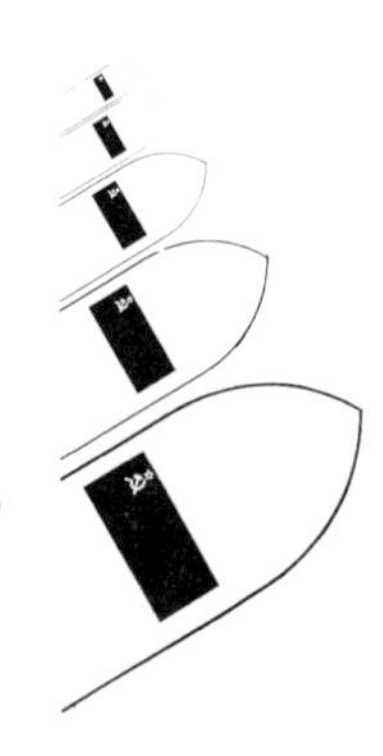

We shall first attempt to estimate the range and the angle of trajectory of the Russian missile based on the data given in Table 1. The Chinese missile will be studied later. For convenience, we plot the four corners of the impact zone in the xy-plane (Fig. 1) even though they actually lie on a nonplanar surface. These points apparently arise from range and direction errors in the missile flight. For instance, impact at P_1 or P_2 indicates that the missile landed short of the desired target, and impact at P_3 or P_4 indicates that it overshot. Since the missiles are not generally too high above the earth's surface they are still under the influence of gravity. Thus we may use two opposite sides of the impact zone to form separate planes containing the center, 0, of the earth to locate the launch site. Let plane *I* contain P_1, P_4 and 0, and plane II contain P_2, P_3, and 0 (Fig. 2). These two planes intersect in a line through the point 0. The line of intersection again intersects the earth's surface at two points, one of which is the launching site and should be inside Russia.

2.2 The Coordinates of Points of Impact

For convenience, we consider the earth a unit sphere. Let the center of the earth coincide with the origin of the Oxyz-system, and let P be a point on the earth's surface (Fig. 3). The equator lies in the xy-plane. The angle θ, measured from the positive x-axis counterclockwise (0°–180°) is called the East Longitude of P, it is called West Longitude if measured clockwise. Similarly, the angle ϕ is called the North Latitude (0°–90°) or South Latitude of P according to whether ϕ is measured upward or downward from the horizontal plane. The ϕ and θ values of the four points of Table 1 are as follows:

Table 2

	P_1	P_2	P_3	P_4
ϕ	10°20′	11°30′	9°10′	8° 5′
θ	189°30′	192° 5′	193°15′	190°40′

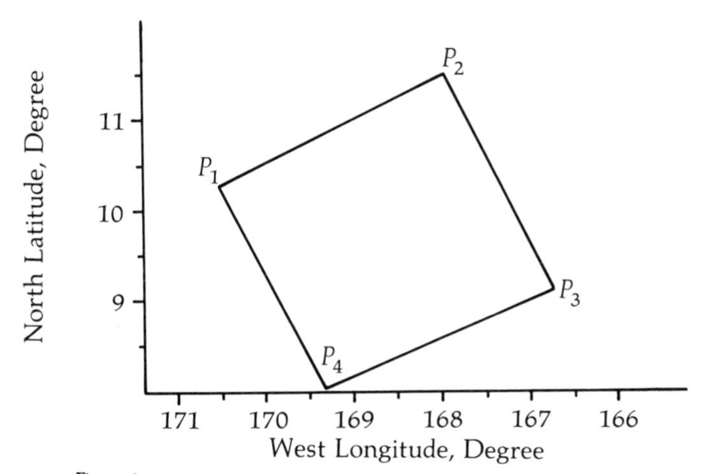

Figure 1

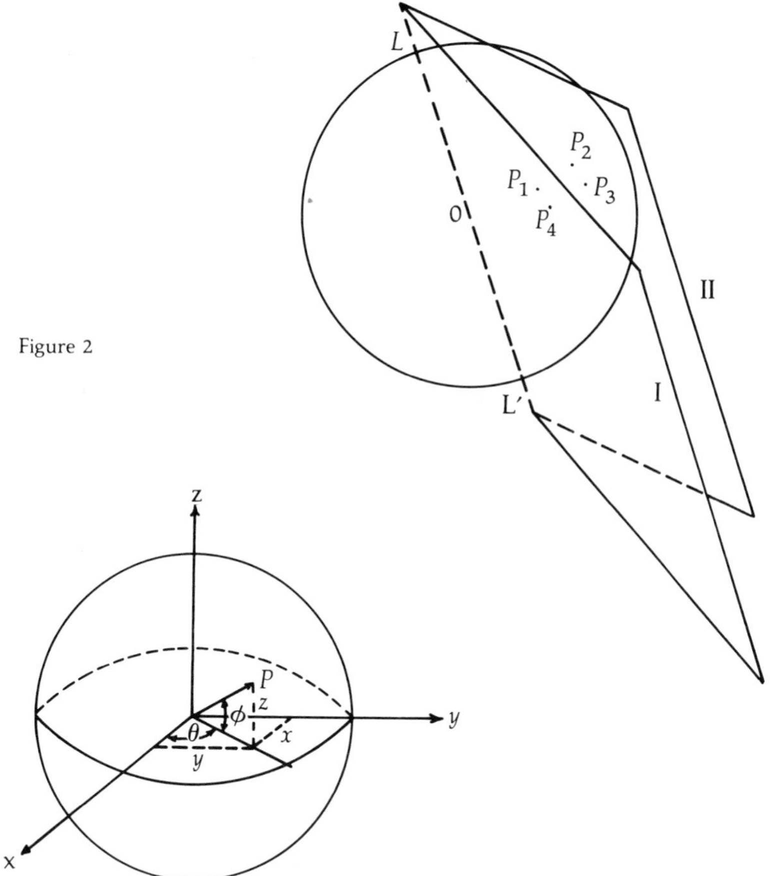

Figure 2

Figure 3

In terms of ϕ and θ the Cartesian coordinates of P are given by

$$x = R\cos\phi\cos\theta,$$

$$y = R\cos\phi\sin\theta, \tag{1}$$

$$z = R\sin\phi. \qquad (R = 1)$$

Using Table 2 and (1) we can calculate the x-, y-, and z-coordinates for each of the four points. The results are contained in the following table.

Table 3

	P_1	P_2	P_3	P_4
x	−0.9702894	−0.9582137	−0.9609479	−0.9729566
y	−0.162371	−0.205136	−0.2262728	−0.183258
z	0.179375	0.199368	0.159307	0.140613

Exercise

1. Assume that $R = 1$. Calculate the Cartesian coordinates of P_4 by using (1) and compare your answers with Table 3.

2.3 A Scalar Triple Product Application

Before we get into any further calculations we recall an important result from vector algebra: if three vectors $\vec{OS}$, $\vec{OT}$, and $\vec{OU}$ are coplanar, then their scalar triple product $\vec{OS} \cdot (\vec{OT} \times \vec{OU})$ is zero. When this relationship is applied on plane I we have (see Fig. 2)

$$\vec{OL} \cdot (\vec{OP}_4 \times \vec{OP}_1) = 0. \tag{2}$$

In determinant form, we have

$$\begin{vmatrix} x & y & z \\ x_4 & y_4 & z_4 \\ x_1 & y_1 & z_1 \end{vmatrix} = 0. \tag{3}$$

Substituting the values from columns under P_4 and P_1 of Table 3 into the second and third rows of the determinant, respectively, and expanding, we obtain

$$-0.0100405x + 0.0380887y - 0.0198333z = 0. \tag{4}$$

Likewise, for plane II we have

$$\vec{OL} \cdot (\vec{OP}_3 \times \vec{OP}_2) = 0, \tag{5}$$

and the corresponding equation in Cartesian coordinates is

$$-0.0124327x + 0.0389321y - 0.0196969z = 0. \tag{6}$$

Since the point L is on the surface of the earth, we must have

$$x^2 + y^2 + z^2 = 1. \tag{7}$$

The solutions of the system consisting of (4), (6), and (7) are found to be

$$\begin{aligned} x &= \pm 0.2227009, \\ y &= \pm 0.4957949, \\ z &= \pm 0.8393995. \end{aligned} \tag{8}$$

Now we must determine the signs of these coordinates. (Do you know how many solutions we may have?) However, a careful examination of the Soviet territory on a globe suggests that we only need to consider positive values of all these coordinates since the country is largely situated in the first octant of the *Oxyz*-system.

Therefore, we have $L = (0.2227009, 0.4957949, 0.8393995)$.

Exercises

2. Verify (4).

3. Verify (6).

4. Solve the system consisting of (4), (6), and (7), and compare your answers with (8).

To determine the test site we must first find out the values of ϕ and θ for L. By using (1)

$$\phi = \sin^{-1} z = \sin^{-1} 0.8393995 = 57°04'36'',$$

and

$$\theta = \sin^{-1}(y/\cos\phi) = \sin^{-1}(0.4957949/0.5435149)$$

$$= 65°48'41''.$$

Thus, the test site is at 57°04′36″ N. Latitude and 65°48′41″ E. Longitude, which is approximately 370 kilometers northeast of Sverdlovsk.

2.4 The Range

We shall define the range of a missile flight as the distance measured along a circular arc between the point of takeoff and the point of landing. From Fig. 4,

$$S = R\,(2\gamma) \qquad (\gamma \text{ is in radian measure})$$
$$= 2\,R \sin^{-1}(d/(2R)). \tag{9}$$

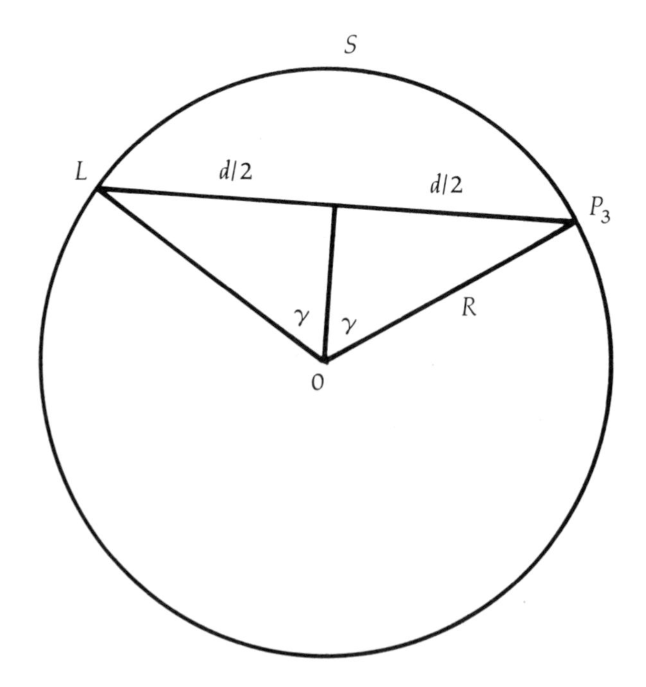

Figure 4

We may use any of the four landing points to calculate an approximate value for S. Suppose that we choose P_3, then

$$d = \left[(x_3 - x)^2 + (y_3 - y)^2 + (z_3 - z)^2 \right]^{1/2} R$$

$$= \left[(-0.9609479-0.2227009)^2 + (-0.2262728-0.4957949)^2 + (0.159307-0.8393995)^2 \right]^{1/2} R = 1.5443224\, R$$

and

$$S = 2\, R \sin^{-1} (1.5443224/2)$$

$$= 2\,(6371) \sin^{-1} 0.7721612 \qquad \text{(use } R = 6371 \text{ km.)}$$

$$= 11{,}241 \text{ km.} \qquad \text{(6985 miles).}$$

2.5 The Direction

The term 'direction' is relative. It refers to the acute angle λ between the equatorial (horizontal) plane E_p and the plane of the missile's trajectory. This angle can be approximated by using either plane I and E_p, or plane II and E_p. From vector algebra, the angle α between two intersecting planes

$A_1x + B_1y + C_1z = D_1$ and $A_2x + B_2y + C_2z = D_2$ is given by

$$\cos \alpha = \frac{A_1A_2 + B_1B_2 + C_1C_2}{(A_1^2 + B_1^2 + C_1^2)^{1/2}\,(A_2^2 + B_2^2 + C_2^2)^{1/2}}, \tag{10}$$

Now suppose we use plane I and E_p whose equation is simply $0x + 0y + z = 0$. Then, using (4) in (10) we get

$$\lambda = \cos^{-1} \frac{0.0198333}{(0.010045^2 + (-0.0380887)^2 + 0.0198333^2)^{1/2}}$$

$$= 63°16'.$$

It is interesting to notice that this angle is quite close to that (65°) of the first Sputnik launched in 1957.

Exercises

5. Calculate the range S of the Russian missile by using L and P_4 and compare your answer with the result of 2.5.
6. Determine the angle between plane II and the horizontal plane E_p.

2.6 The Aiming Accuracy

Since the real target position and the original angle of the plane of trajectory were not known, it is difficult to calculate the angle deflections. However, we may consider that the intended plane of trajectory lies midway between plane I and plane II. The angle ϵ between these two planes is, by (4), (6), and (10)

$$\epsilon =$$

$$\cos^{-1} \frac{-0.0100405\,(-0.0124327) + (0.0380887)\,(0.0389321) + (-0.0198333)\,(-0.0196969)}{\left[0.0100405^2 + 0.0380887^2 + 0.0198333^2\right]^{1/2} \left[0.0124327^2 + 0.0389321^2 + 0.0196969^2\right]^{1/2}}$$

$$= \cos^{-1} 0.9987954 = 2°48'45'',$$

indicating that the angle deflection on either side of the intended plane of trajectory is less than 1.5°.

Exercise

7. In October, 1962, the Soviets test-fired their ICBM into the Pacific. The impact zone was given as follows:

	P_1	P_2	P_3	P_4
North Latitude	6°52′	5°10′	4°21′30″	6°7′
West Longitude	165°25′	164°36′30″	166°16′30″	167°8′

following the steps and procedures given in 2.2, 2.3, 2.4, and 2.5 find:

(1) ϕ and θ for L,

(2) S,

(3) λ,

(4) ϵ.

3. The Chinese Missile

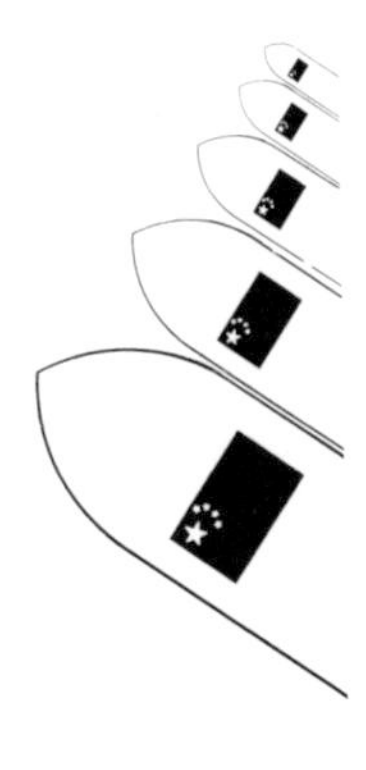

You must have noticed that in the introduction of this module, there were two types of impact zones mentioned. One is nearly rectangular and the other is circular. In fact, after 1970 only the second type was always used by the Soviets to inform the world of their missile tests. The Chinese did the same in 1980. Unfortunately, this change presents some difficulties in analyzing the problem because two of the three conditions used to determine the test site were absent. It is quite possible that the original rectangular zone had been inscribed in a circular zone but we have no way of knowing it. (You may disagree.)

At this point, the best we can do is to assume that the Chinese test site L_c is near the Nop Nur (41° North Latitude, 90° East Longitude) complex as many scientists and militarists thought it to be, and that the center (7° South Latitude, 171°33′ East Longitude) of the circle represents the target, P_c. Then the distance L_c and P_c can be calculated.

	Nop Nur, L_c	Target, P_c
ϕ	41°	-7°
θ	90°	171°33′
x	0	-0.9817716
y	0.7547096	0.145851
z	0.656059	-0.1218693

$$d(L_cP_c) = \left[(-0.9817716)^2 + (0.141851 - 0.7547096)^2 + (-0.1218693 - 0.656059)^2\right]^{1/2}$$

$$= 1.3897042.$$

Hence the range is

$$S = 2R \sin^{-1}(1.3897042/2)$$

$$= (6371)\,(1.5364284)$$

$$= 9788.59 \text{ km}, \qquad (6080 \text{ miles})$$

and this figure is in close agreement with that reported on May 19, 1980 by *The New York Times.*

4. Answers to Exercises

2.

$$\begin{vmatrix} x & y & z \\ x_4 & y_4 & z_4 \\ x_1 & y_1 & z_1 \end{vmatrix} = \begin{vmatrix} x & y & z \\ -0.9729566 & -0.183258 & 0.140613 \\ -0.9702894 & -0.162371 & 0.179375 \end{vmatrix}$$

$$= x \begin{vmatrix} -0.183258 & 0.140613 \\ -0.162371 & 0.179375 \end{vmatrix} + y \begin{vmatrix} 0.140613 & -0.9729566 \\ 0.179375 & -0.9702894 \end{vmatrix} + z \begin{vmatrix} -0.9729566 & -0.183258 \\ -0.9702894 & -0.162371 \end{vmatrix}$$

$$= x\,(-0.0328719 + 0.0228314) + y\,(-0.1364353 + 0.174524) + z\,(0.1579799 - 0.1778132)$$

$$= -0.0100405x + 0.0380887y - 0.0198333z = 0.$$

3.

$$\begin{vmatrix} x & y & z \\ x_3 & y_3 & z_3 \\ x_2 & y_2 & z_2 \end{vmatrix} = \begin{vmatrix} x & y & z \\ -0.9609479 & -0.2262728 & 0.159307 \\ -0.9582137 & -0.2051316 & 0.199368 \end{vmatrix}$$

$$= x \begin{vmatrix} -0.2262728 & 0.159307 \\ -0.2051316 & 0.199368 \end{vmatrix} + y \begin{vmatrix} 0.159307 & -0.9609479 \\ 0.199368 & -0.9582137 \end{vmatrix} + z \begin{vmatrix} -0.9609479 & -0.2262728 \\ -0.9582137 & -0.2051316 \end{vmatrix}$$

$$= x\,(-0.0451115 + 0.0326788) + y\,(-0.1526501 + 0.1915822) + z\,(0.1971207 - 0.2168176)$$

$$= -0.0124327x + 0.0389321y - 0.0196969z = 0.$$

4. From **(6)**,

$$0.0389321y = 0.0124327x + 0.0196969z$$

(a) $y = 0.3193431\,x + 0.5059295\,z.$

Substituting in **(4)** we have

$$-0.0100405\,x + (0.0380887)\,(0.3193431x + 0.5059295z) -$$

$$0.0198333z = 0,$$

and after simplifying we get

(b) $x = 0.2653099\,z.$

Thus **(a)** simplifies to

$$y = 0.3193431\,(0.2653099\,z) + 0.5059295\,z$$

(c) $\quad = 0.0847248\,z + 0.5059295\,z = 0.5906543\,z.$

Now substituting **(b)** and **(c)** in **(7)**,

$$(0.2653099z)^2 + (0.5906543z)^2 + z^2 = 1,$$

$$1.4192618\,z^2 = 1,$$

$$z = \pm 0.8393995.$$

Hence, $\qquad x = \pm 0.2227009,$

and $\qquad y = \pm 0.4957949.$

5. $d = [\,(-0.9729566-0.2227009)^2 + (-0.183258-0.4957949)^2 +$

$$(0.140613 - 0.8393995)^2\,]^{1/2}\,R = 1.5424039\,R.$$

$S = 2R\sin^{-1}(1.5424039/2) = 2\,(6317)\sin^{-1}0.771202 = 11{,}222$ km.

6. $\lambda = \cos^{-1}\dfrac{0.0196969}{(0.0124327^2 + (-0.0389321)^2 + (0.0196969)^2)^{1/2}}$

$= 64°16'05''.$

7.

	P_1	P_2	P_3	P_4
ϕ	6°52′	5°10′	4°21′30″	6°7′
θ	194°35′	195°23′30″	193°43′30″	192°52′
x	−0.9608407	−0.9602167	−0.9686366	−0.9693408
y	−0.2499818	−0.2643375	−0.2365760	−0.2214153
z	−0.1195593	0.0900532	0.0759939	0.1065533

The equation of the plane containing 0, P_3, and P_4 is

$$\begin{vmatrix} x & y & z \\ x_4 & y_4 & z_4 \\ x_3 & y_3 & z_3 \end{vmatrix} = \begin{vmatrix} x & y & z \\ -0.9693408 & -0.2214153 & 0.1065533 \\ -0.9686366 & -0.2365760 & 0.0759939 \end{vmatrix}$$

$$= x \begin{vmatrix} -0.2214153 & 0.1065533 \\ -0.2365760 & 0.0759939 \end{vmatrix} +$$

$$y \begin{vmatrix} 0.1065533 & -0.9693408 \\ 0.0759939 & -0.9686366 \end{vmatrix} +$$

$$z \begin{vmatrix} -0.9693408 & -0.2214153 \\ -0.9686366 & -0.2365760 \end{vmatrix}$$

$$= x(-0.0168262 + 0.025208) + y(-0.1032114 + 0.073664) + z(0.2293228 - 0.214471)$$

(a)

$$= 0.0083818\,x - 0.0295474\,y + 0.0148518\,z = 0.$$

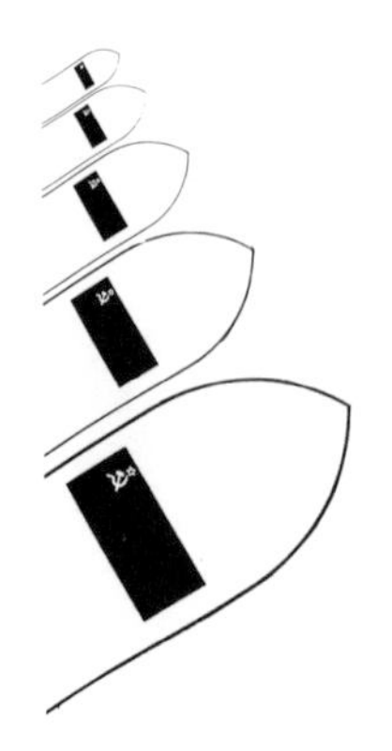

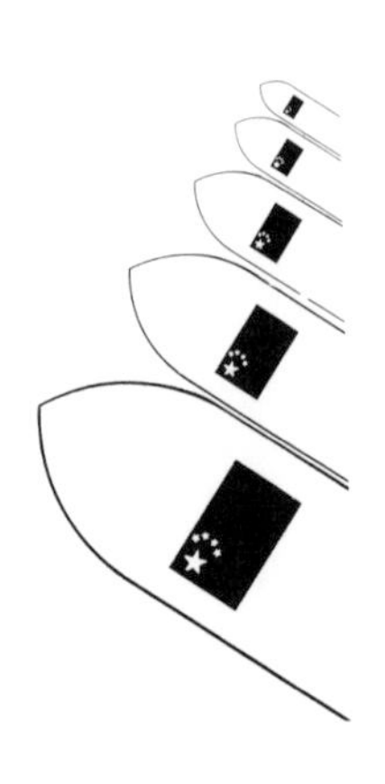

The equation of the plane containing 0, P_1, and P_2 is

$$\begin{vmatrix} x & y & z \\ x_1 & y_1 & z_1 \\ x_2 & y_2 & z_2 \end{vmatrix} = x \begin{vmatrix} -0.2499818 & 0.1195593 \\ -0.2643375 & 0.0900532 \end{vmatrix} + y \begin{vmatrix} 0.1195593 & -0.9608407 \\ 0.0900532 & -0.9602167 \end{vmatrix} + z \begin{vmatrix} -0.9608407 & -0.2499818 \\ -0.9602167 & -0.2643375 \end{vmatrix}$$

$$= x(-0.0225117 + 0.031604) + y(-0.1148028 - +0.0865268) + z(0.2539862 - 0.2400367)$$

$$= 0.0090923x - 0.028276y + 0.0139495z = 0.$$

From **(b)**,

$$y = \frac{0.0090923\,x + 0.0139495\,z}{0.028276}$$

$$= 0.3215554\,x + 0.4933336\,z.$$

Substituting in **(a)** we have

$$0.0083818x - 0.0295474(0.3215554x + 0.4933336z) + 0.0148518z = 0.$$

So

$$x = \frac{0.0002751z}{0.0011193} = 0.2457786z,$$

and

$$y = 0.3215554(0.2457786)z + 0.4933336z$$

$$= 0.572365z.$$

Now using (7),

$$\left[(0.2457786)^2 + (0.572365)^2 + 1\right] z^2 = 1,$$

$$1.3880088\, z^2 = 1,$$

$$z = \pm 0.8487971.$$

Hence, $\quad y = \pm 0.4858218,$

and $\quad x = \pm 0.2086162.$

$$\phi = \sin^{-1}\ 0.8487971 = 58°04'52''.$$

$$\theta = \sin^{-1} \frac{0.4858218}{0.5287187} = \sin^{-1}\ 0.9188663$$

$$= 66°45'39''.$$

$$d = \left[(-0.9602167 - 0.2086162)^2 + (-0.2643375 - 0.4858218)^2 + (0.0900532 - 0.8487971)^2\right]^{1/2} R$$

$$= 1.5835933\ R.$$

$$S = 2\ R \sin^{-1}\ (1.5825933/2) = 2\ (6371)(0.9129269)$$

$$= 11633 \text{ km}.$$

We use **(a)** to calculate λ.

$$\lambda = \cos^{-1} \frac{0.0148518}{(0.008318^2 + (-0.0295474)^2 + 0.0148518^2)^{1/2}}$$

$$= \cos^{-1}\ 0.4353365 = 64°11'36''.$$

$$\epsilon =$$

$$\cos^{-1} \frac{0.0083818(0.0090923)+(0.0295474)(0.028276)+(0.0148518)(0.0139495)}{\left[0.0083818^2+0.0295474^2+0.0148518^2\right]^{1/2}\left[0.0090923^2+0.028276^2+0.0139495^2\right]^{1/2}}$$

$$= \cos^{-1} 0.9994733 = 1°52'.$$

14

UMAP

Modules in Undergraduate Mathematics and its Applications

Module 662

Computer Implementation of Matrix Computations and Row Transformations on Matrices With Application to Solving Systems of Linear Equations

Wendell L. Motter

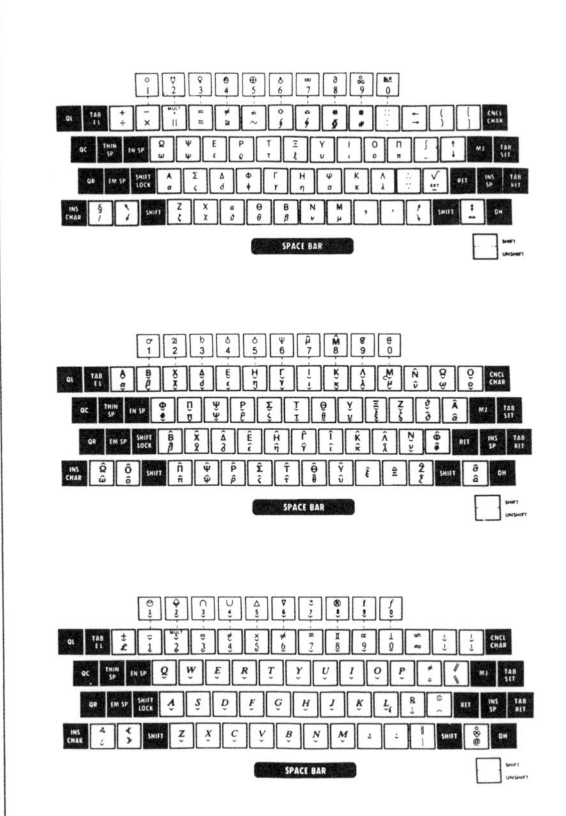

Published in cooperation with the Society for Industrial and Applied Mathematics, the Mathematical Association of America, the National Council of Teachers of Mathematics, the American Mathematical Association of Two-Year Colleges, and The Institute of Management Sciences.

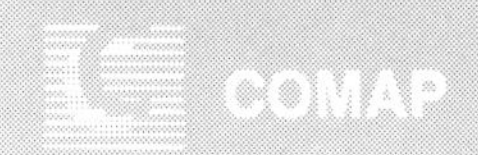

INTERMODULAR DESCRIPTION SHEET: UNIT 662

TITLE: COMPUTER IMPLEMENTATION OF MATRIX COMPUTATIONS AND ROW TRANSFORMATIONS ON MATRICES WITH APPLICATION TO SOLVING SYSTEMS OF LINEAR EQUATIONS

AUTHOR: Wendell L. Motter
Department of Mathematics
Florida A and M University
Tallahassee, FL 32307

MATH FIELD: Computer Science

APPLICATION FIELD: Computer Science

TARGET AUDIENCE: Students in Computer Science

ABSTRACT: This module gives a comprehensive introduction to the use of a computer to perform a variety of matrix operations and calculations. Computer programming techniques for the input/output of matrices, matrix-arithmetic operations, computations of powers of square matrices, and the implementation of elementary row transformations on matrices are discussed thoroughly. A "generic" version of the BASIC programming language is used throughout and the language features employed are available on most micro, mini, and macro computer systems.

PREREQUISITES: Matrix conventions and notations; matrix arithmetic operations. Gaussian Elimination Algorithm; row transformations on matrices.

Computer Implementation of Matrix Computations and Row Transformations on Matrices With Application to Solving Systems of Linear Equations

Wendell L. Motter
Department of Mathematics
Florida A and M University
Tallahassee, FL 32307

Table of Contents

1. INTRODUCTION 1
2. INPUT AND OUTPUT OF MATRICES USING BASIC 3
3. PRINTING ENTRIES OF A MATRIX ROW ALONG ONE LINE 18
    3.1 Computations Using Matrices 22
    3.2 Rounding Off Matrix Entries to a Specified Precision 30
    3.3 Computer Implementation of Matrix-Arithmetic Operations 28
    3.4 Application of Computer-Matrix-Multiplication to Dietary Analysis 35
4. COMPUTER IMPLEMENTATION OF ELEMENTARY ROW TRANSFORMATIONS ON MATRICES 40
5. COMPUTER IMPLEMENTATION OF THE GAUSSIAN ELIMINATION ALGORITHM 52
6. BIBLIOGRAPHY 61
7. ANSWERS TO SELECTED EXERCISES 63

Modules and Monographs in Undergraduate Mathematics and its Applications Project (UMAP)

The goal of UMAP was to develop, through a community of users and developers, a system of instructional modules in undergraduate mathematics and its applications to be used to supplement existing courses and from which complete courses may eventually be built.

The Project was guided by a National Advisory Board of mathematicians, scientists, and educators. UMAP was funded by a grant from the National Science Foundation and is now supported by the Consortium for Mathematics and Its Applications, Inc. (COMAP), a nonprofit corporation engaged in research and development in mathematics education.

This material was prepared with the partial support of National Science Foundation Grant No. SPE8304192. Recommendations expressed are those of the author and do not necessarily reflect the views of the NSF or the copyright holder.

1. Introduction

This unit gives a comprehensive introduction to the use of a computer to perform a variety of matrix operations and calculations. Computer programming techniques for the input/output of matrices, matrix-arithmetic operations, computation of powers of square matrices, and the implementation of elementary row transformations on matrices are discussed thoroughly. A "generic" version of the BASIC programming language is used throughout and the language features employed are available on most micro, mini, and macro computer systems.

Section 5 develops a complete program that enables a student to "interactively" use a computer to employ various row transformations on a matrix in a step-by-step fashion. This program can be used to perform the various steps (designated one at a time by the program user) of a Gaussian Elimination strategy that results in a "row-echelon" matrix which is row-equivalent to the starting matrix. Alternatively, this computer program can be used to carry out the steps of the simplex algorithm applied to the initial simplex tableau of a linear programming problem expressed in standard form.

Each major computational technique is used to analyze significant and interesting real-world applications. For example, the method of finding the sum of the entries in a matrix row is applied to household budgets and record keeping. The computer program for matrix multiplication is used to implement a model of dietary analysis involving food compositions and quantities consumed. Of course, the major application is the computer implementation of the Gaussian Elimination Algorithm for solving systems of linear equations with real-number variables.

This unit can serve as an important supplement to courses in Finite Mathematics and Elementary Linear Algebra, especially when it is desirable to introduce computer calculations using matrices without requiring students to have prior computer programming experience. Moreover, the comprehensive program for interactive usage of a computer to carry out specified row transformations on a matrix should enable a student to more easily learn the Gaussian Elimination Algorithm (or Simplex Method) since the tedious and time-consuming arithmetic calculations will be done "instantaneously" — by the computer.

2. Input and Output of Matrices Using BASIC

In this section, we will discuss how most computer systems (including the "home" computers like TRS80, Apple, etc.) can be programmed to process matrices with numerical entries. Matrices are usually classified as two-dimensional arrays in many discussions of the features of BASIC concerned with them. Tables consisting of rows and columns of numbers is a very general way of understanding matrices as a computer data structure.

Computer processing of matrices is set up to organize the storage and calculations with matrices in essentially the same way that we treat them mathematically. Recall that the entries of any matrix are organized into rows and columns so that the entry in row *I* and column *K* of a matrix *A* can be represented by the abstract, mathematical notation a_{Ik} using the subscripts *I* for the row location and *K* for the column location. For example, any 3×4 matrix *A* can be represented as:

$$\begin{matrix} a_{11} & a_{12} & a_{13} & a_{14} \\ a_{21} & a_{22} & a_{23} & a_{24} \\ a_{31} & a_{32} & a_{33} & a_{34} \end{matrix}$$

In computer notation, a_{Ik} becomes A(I,K) since the typing (using a computer terminal keyboard) or printing (using a line printer) usually must be done without using subscripts (or superscripts). Using computer notation, a 3 × 4 matrix A would be represented as:

```
A(1,1)   A(1,2)   A(1,3)   A(1,4)
A(2,1)   A(2,2)   A(2,3)   A(2,4)
A(3,1)   A(3,2)   A(3,3)   A(3,4)
```

Observe also that Row *I* is *A*(*I*,1) *A*(*I*,2) *A*(*I*,3) *A*(*I*,4) where *I* can be 1, 2, or 3 in this example.

In computer memory there are memory slots referenced by *A*(1,1), *A*(1,2), etc. which store the actual values of the matrix entries. The numeric values initially used as the entries of the matrix are called data values or simply data. BASIC has several programming techniques for assigning data values to a specified matrix. Let us consider first a method that is straightforward and easy to understand.

Suppose we have the 3×4 matrix A =

$$\begin{bmatrix} 1 & 1.2 & 1.3 & 1.056 \\ 2 & 2.2 & 2.3 & 2.056 \\ 3 & 3.2 & 3.3 & 3.056 \end{bmatrix}$$

The following program segment setups computer memory slots $A(1,1)$, $A(1,2)$,... $A(3,4)$ and picks up the corresponding data values 1, 1.2, ..., 3.056 for all the matrix entries.

Example 2.1

```
100  DIM A(3,4)
120  READ A(1,1), A(1,2), A(1,3), A(1,4)
122  READ A(2,1), A(2,2), A(2,3), A(2,4)
124  READ A(3,1), A(3,2), A(3,3), A(3,4)
195  REM DATA FOR MATRIX IS ARRANGED IN ROWS
196  REM THE ROW ENTRIES ARE SEPARATED BY
     COMMAS
200  DATA 1, 1.2, 1.3, 1.056
202  DATA 2, 2.2, 2.3, 2.056
204  DATA 3, 3.2, 3.3, 3.056
300  END
```

The BASIC statement (coded instruction) DIM A(3,4) is given the line number 100 and is therefore the first instruction executed by the computer as it proceeds to perform all the instructions sequentially according to the numerical order of their associated line numbers. DIM A(3,4) stands for DIMENSION A(3,4) which organizes the computer memory setup for the matrix A in correspondence with the shape of A, i.e., three rows and four columns. As a general rule, BASIC programs involving matrices require that DIM statements be at the beginning of the program and that all matrices being used have a DIM statement which specifies their shape.

The instruction (line number 120) READ A(1,2), A(1,2), A(1,3), A(1,4) works by assigning to the variables A(1,2), A(1,2), A(1,3), A(1,4), representing row 1 of A, the numbers (in this example 1, 1.2, 1.3, 1.056) provided in the first DATA statement (line number 200). Similarly, when the READ instruction given by line 122 is executed, then the variables $A(2,1)$, $A(2,2)$, $A(2,3)$, $A(2,4)$ which represent row 2 of A are assigned the numbers 2, 2,1, 2,3, 2.056 provided in the second DATA statement (line 202). The variables representing row 3 of A are assigned numbers likewise by the associated READ and DATA statements given in lines 124 and 204. Before completing our discussion of example

2.1, we should note that the instructions numbered 195 and 196 are nonessential "remark" statements and are used to carry explanatory comments that describe relevant features of the program.

The method illustrated by example 2.1 can be generalized to work for any $m \times n$ matrix by using similar READ and corresponding DATA statements for each row of matrix. For matrices with several entries in each row, two or more READ statements may be needed for each row. If, for example, row 1 of matrix A has nine entries such as 1, 2.2, 3.56, 4, 5.1, 6.2, 7.88, 8.09, 9.0, then you can use instructions like:

```
120  READ A(1,1), A(1,2), A(1,3), A(1,4), A(1,5)
121  READ A(1,6), A(1,7), A(1,8), A(1,9)
     with corresponding DATA statements
200  DATA 1, 2.2, 3.56, 4, 5.1
201  DATA 6.2, 7.88, 8.09, 9.0
```

Variations of this may also work, but as a general rule the number of characters typed for a BASIC program instruction (including the characters of the line number and spaces) should not exceed 72.

Exercise 2.1

Design a program segment similar to example 2.1 to set up and store the matrix A given by

$$\begin{bmatrix} 3.46 & 2 & 5 & 7.01 & 5 & 6.3 \\ 77.5 & 3.6 & 4 & 9 & 100.1 & 8.7 \end{bmatrix}$$

The method for matrix input discussed for example 2.1 becomes cumbersome and impractical for larger matrices. Suppose a matrix A has M rows, the following program segment is an improvement of the previous method. Again, we use A =

$$\begin{bmatrix} 1 & 1.1 & 1.3 & 1.056 \\ 2 & 2.2 & 2.3 & 2.056 \\ 3 & 3.2 & 3.3 & 3.056 \end{bmatrix}$$

Example 2.2

```
100  DIM A(3,4)
105  READ M
110  FOR I = 1 TO M STEP + 1
120  READ A(I,1), A(I,2), A(I,3), A(I,4)
130  NEXT I
```

```
190 REM DATA FOR NUMBER OF ROWS
191 DATA 3
199 REM DATA FOR MATRIX IS ARRANGED IN ROWS
200 DATA 1, 1.1, 1.3, 1.056
202 DATA 2, 2.2, 2.3, 2.056
204 DATA 3, 3.2, 3.3, 3.056
300 END
```

The new concept is the use of a FOR-NEXT loop which in this example is composed of the instructions numbered 110, 120, and 130. This loop makes the instruction READ A(I,1), A(I,2), A(I,3), A(I,4) (which assigns numbers to all row I variables) be executed for the value of $I = 1$ first, then $I = 2$ second, and finally $I = M = 3$. As before, the numbers for the entries in each matrix row are provided by DATA statements for each row.

Let's explain more thoroughly how this program segment works. The statement FOR $I = 1$ TO M STEP +1 specifies I as the "loop-control" variable (or loop index), assigns I the starting value of 1, and sets up the value of M as the last value for I. Note, earlier in the program, M is given the value 3 by the READ M instruction and the associated statement DATA 3. The instruction 130 NEXT I changes the value of I by +1 (the number following the word STEP) each time the loop is executed and thus, the instruction (or instructions) enclosed by the FOR-NEXT statements are repeated until the variable I is changed to a value which exceeds the value of M.

Exercise 2.2

Design a program segment similar to example 2.2 to set up and store the matrix A given by

$$\begin{bmatrix} 1 & 7 & 8 & 6.2 & 90 \\ 2 & 5.03 & 7 & 17 & 10 \\ 3 & 66.7 & 50.5 & 3 & 70 \end{bmatrix}$$

The method of example 2.2 allows for a variable number of matrix rows and thereby simplifies the input of matrices with any number of rows. This method becomes impractical for matrices with a large number of columns, however.

Suppose the variable N represents the number of columns in a matrix A. Consider the following program segment

```
120 FOR J = 1 TO N STEP + 1
122 READ A(I,J)
125 NEXT J
```

This is a FOR-NEXT loop with *J* the loop variable. This makes the variable *J* proceed through the values 1,2,3, ..., *N* and performs the instruction READ A(I,J) for each value. Thus, READ A(I,1) is done when *J* = 1, READ A(I,2) is done when *J* = 2, and so on. Data for each row *I* must be provided in data statements of course, and *N* must be given a data value also.

If we could make this entire loop be repeated for each value of *I*, then we could read in the values for the matrix in a row-by-row fashion. This can be accomplished by placing the instructions for the above loop inside a FOR-NEXT loop that is controlled by the variable *I*. This method is illustrated by the following example.

Example 2.3

```
100  DIM (A3,4)
105  READ M
107  READ N
110  FOR I = 1 TO M STEP + 1 ─────────────┐
120  FOR J = 1 TO N STEP + 1─┐             │
122  READ A(I,J)             │ inner loop  │ outer loop
125  NEXT J ─────────────────┘             │
130  NEXT I ───────────────────────────────┘
190  REM DATA FOR NUMBER OF ROWS
191  DATA 3
192  REM DATA FOR NUMBER OF COLUMNS
193  DATA 4
199  REM DATA FOR MATRIX IS ARRANGED IN ROWS
200  DATA 1, 1.1, 1.3, 1.056
202  DATA 2, 2.2, 2.3, 2.056
204  DATA 3, 3.2, 3.3, 3.056
300  END
```

This accomplishes the same results as the method given in Example 2.2. Initially, *I* = 1 by the instruction at line 110, and while *I* = 1, the inner loop makes *J* take on the values 1,2,3, ..., *N* and executes READ A(1,J) for each *J* value. Thus, row 1 will be input. Next, *I* changes to its next value, namely *I* = 2. Again, the innerloop is repeated and *J* takes on the values 1,2,3, ..., *N* and READ *A*(2,J) is executed for each *J* value so that row 2 will be input. This repetitive procedure continues for each value of *I*; *I* = *M* is the ending value.

The method illustrated by example 2.3 is a very general programming technique in BASIC and can be used to input data for any $M \times N$ matrix. The data for *M*,*N*, and the matrix row entries can be provided using appropriate DATA statements similar to those in example 2.3 (see instructions numbered 191, 193, 200,

202, and 204). No other changes in the programming instructions would be necessary. This is what makes the design and use of computer programs so worthwhile, namely, that the same program can be applicable to many problems of a similar nature by simply changing the data and making the DIM statements match the shapes of the matrices.

Exercise 2.3

Design a program segment similar to example 2.3 to set up and store the matrix A =

$$\begin{bmatrix} 20 & 1.2 & 7.5 \\ 15 & 3 & 66.7 \\ 8 & 5 & 5 \\ 9.9 & 1 & 6.0 \end{bmatrix}$$

So far we have discussed only the input of matrices and introduced the relevant features of the BASIC programming language to use a computer for this purpose. Next, we consider some methods for creating matrix output. This usually means the printed results that appear on a computer terminal screen when a program is "run" on a computer. Of course, if a printer is available, then output is created by the printed results on some kind of paper. This is usually called hard copy output.

To print out the entries of a matrix A in a row by row fashion, we can employ a strategy of nested loops (inner and outer loops controlled by the column index J and row index I respectively) as in example 2.3 where we needed to input matrix entries. The following program segment can be used.

```
150  FOR I = 1 TO M STEP + 1
152  FOR J = 1 TO N STEP + 1
154  PRINT A(I,J),
156  NEXT J
157  PRINT " "
158  PRINT " "
160  NEXT I
```

This prints out the values $A(I,1)$ $A(I,2) \ldots A(I,N)$ along the same line (or subsequent lines depending on the number of row entries) because of the PRINT A(I,J), being executed for $J = 1,2,\ldots, N$. Then a blank line is printed by instructions numbered 157 and 158 and I continues through all the values $1,2,\ldots, M$.

When this segment of instructions is added to the instructions of example 2.3 (since data for the matrix must be input first), then the following kind of output is printed on the terminal screen—this will vary depending on what computer system is being used due to features such as the number of columns for terminal output, the "field-width" allowed for each number in a matrix row, etc.

```
            10        20        30        40— columns
1234567890123456789012345678901234567890 — labeling of
                                           column

1               1.1             1.3  ]
1.056                                ]   rows of matrix
2               2.2             2.3  ]   A (actual output
2.056                                ]   seen on terminal
3               3.2             3.3  ]   screen)
3.056
```

Figure 2.4

Note: matrix A equals

$$\begin{bmatrix} 1 & 1.1 & 1.3 & 1.056 \\ 2 & 2.2 & 2.3 & 2.056 \\ 3 & 3.2 & 3.3 & 3.056 \end{bmatrix}$$

The output shown in Fig. 2.4 was created using an Apple II-plus computer which has 40 columns per line of printed output on its terminal screen. This demonstrates how the row entries are printed (starting in the left-hand column) within fields of uniform width—in this case 16 columns. The comma in the PRINT instruction creates such uniform-width fields; however, the number of columns composing these will vary according to the type of computer (16 for TRS-80 and Apple, 14 for PDP-8, PDP-11, etc.). Fig. 2.4 illustrates also that regardless of how many columns are available per line of printed output, whenever all of the entries of a matrix row cannot fit along one printed line, the computer will use subsequent lines to finish printing an entire matrix row. Consider the following complete program.

Example 2.5

```
100 DIM A(3,5)
105 READ M
107 READ N
110 FOR I = 1 TO M STEP + 1
120 FOR J = 1 TO N STEP + 1
122 READ A(I,J)
```

```
125  NEXT J
130  NEXT I
150  FOR I = 1 TO M STEP + 1
151  PRINT "ROW"; I
152  FOR J = 1 TO N STEP + 1
154  PRINT A(I,J),
156  NEXT J
157  PRINT " "
158  PRINT " "
160  NEXT I
190  REM DATA FOR NUMBER OF ROWS
191  DATA 3
192  REM DATA FOR NUMBER OF COLUMNS
193  DATA 5
199  REM DATA FOR MATRIX IS ARRANGED IN ROWS
200  DATA 1, 1.1, 1.3, 1.056, 1.5
202  DATA 2, 2.2, 2.3, 2.056, 2.5
204  DATA 3, 3.2, 3.3, 3.056, 3.5
300  END
```

When the program of example 2.5 is RUN on a computer (with 40 columns for printed output on the terminal screen), the following output is produced.

```
1234567890123456789012345678901234567890  ←labeling of
                                             columns
ROW1
1               1.1             1.3
1.056           1.5
ROW2                                         rows of matrix
2               2.2             2.3          A (actual output
2.056           2.5                          seen on terminal
ROW3                                         screen)
3               3.2             3.3
3.056           3.5
```

Here A has five entries in each row, and these are printed (as shown) along two lines of output for each row. The additional instruction 151 PRINT "ROW"; I creates the phrases ROW 1, ROW 2, and ROW 3 labeling the three rows when I (because of the outer FOR-NEXT loop) takes on the values 1, 2, and 3.

Exercise 2.5

(a) Design a program similar to Example 2.5 to input and print out the matrix A =

$$\begin{bmatrix} 1 & 1.1 & 1.2 & 1.3 & 1.4 & 1.5 & 1.6 \\ 2 & 2.1 & 2.2 & 2.3 & 2.4 & 2.5 & 2.6 \end{bmatrix}$$

(b) Use a computer to process your BASIC program and examine the output that is generated. Some procedures to follow when using a computer are:

(1) Each BASIC instruction must start with a line number. It is best to use a sequence of line numbers like 100, 105, 110, etc. so additional instructions can be inserted later if changes in your program are made.

(2) After a BASIC instruction is typed, the RETURN (or Enter on TRS-80) key is pressed before typing the next line of BASIC code.

(3) Typing errors that you notice (before you finish an instruction) can be "erased" by moving the "cursor" back to the left and retyping the necessary code.

(4) The complete program can be displayed on the screen by typing the system command LIST.

(5) If all of the program cannot be displayed on the terminal screen, use commands like LIST 100-200 which, for example, displays program statements from line 100 through line 200.

(6) If there is an error in a particular line, simply retype the same line number and the corrected version of that line. This will replace the previous version of the line.

(7) To execute the program, type RUN and press the RETURN key.

(8) If the computer finds something wrong, when it is executing a program it will usually print out the line numbers where there are errors and some description of the errors involved.

3. Printing Entries of a Matrix Row Along One Line

Frequently, it is desirable to print all entries of each row so that these fit along only one line of printed output. In other words, the matrix printed on the terminal screen (or pages of a printer) looks like the matrix written on paper. This may not even be possible unless the matrix has a "small" number of columns and the matrix entries consist of numbers of limited magnitude. Moreover, the strategies to accomplish this depend somewhat on the kind of computer used because the primary consideration will be the number of columns used by the computer for printed output.

Our discussion will be based on computers which have 40 columns for printed output on their terminal screens. Two widely used computers of this nature are the Apple II+ and TRS 80. The techniques developed can easily be extended to computers which use more columns for printed output. Suppose the data for the matrix entries and the computations carried out by the computer will involve numbers that require at most five columns for printing. If the numbers are rounded to the nearest 1/10, this is the set of decimal numbers x, with x satisfying $-99.9 \leq x \leq 999.9$; if the numbers are rounded to the nearest 1/100, this is the set of decimal numbers x, with x satisfying $-9.99 \leq x \leq 99.99$, etc. Note that the "-" sign and decimal point will each use one column. When only integers (n) are used, n must satisfy the condition that $-9999 \leq n \leq 99999$. A technique for rounding-off all matrix entries to any desired precision (nearest tenth, hundredth, etc.) will be discussed in Section 3.1.

Now if we allow for a space (one column) between each printed number, we need to allocate six columns (a field of length 6) for each number printed in a row. The best use of the 40 columns available for printed output is made by printing the numbers so that the first starts in column 1, the second in column 7, the third in column 13, etc. This is illustrated by the following programming instructions and the corresponding output generated as shown in Fig. 2.6. Here, matrix A =

$$\begin{bmatrix} 1 & 22.13 & 15 & 44.56 \\ 2 & -3.45 & 1.78 & 3456 \end{bmatrix}$$

is already stored and M is the number of rows.

```
150  FOR I = 1 TO M
151  PRINT "ROW"; I
155  PRINT TAB( 1);A(I,1); TAB( 7);A(I,2); TAB( 13);A(I,3);
     TAB( 19);A(I,4);
157  PRINT "  "
158  PRINT "  "
160  NEXT I
```

```
             10        20        30        40— columns
1234567890123456789012345678901234567890 — labeling of
                                            columns
ROW1                                   ]  rows of matrix
1     22.33  15    44.56               |  A (actual output
ROW2                                   |  seen on terminal
2     -3.45  1.78  3456                ]  screen)
```

Figure 2.6

The key idea is the use of the TAB( ) function which causes the computer printing process to work like the tab settings on a typewriter and to move so the printing of a number (or numerical value of a variable) starts in the column specified by each value used for TAB( ).

A more general approach must allow for the number of columns to be variable and not become unmanageable as does the type of programming instruction given by line number 155 when the number of columns increases. The following program segment generalizes the process of using the TAB( ) function to the situation with a variable, *N*, specifying the number of columns.

Example 2.7

```
150  FOR I = 1 TO M
151  PRINT "ROW"; I
152  FOR J = 1 TO N
155  REM SET FIELD-LENGTH FOR TAB SPACING
156  LET T = 6 * (J - 1) + 1
157  PRINT TAB( T);A(I,J);
158  NEXT J
159  PRINT "  "
160  PRINT "  "
165  NEXT I
```

Here TAB(T) becomes TAB(1), TAB(7), TAB(13), TAB(19) for *J* = 1,2,3,4 and this produces exactly the same matrix output as shown in Fig. 2.6. Observe that with field length 6, *N* can be at most 6, because the computer can print within 40 columns at most six numbers that require six columns each.

When the numbers are to be printed within nine columns and with a space between each number in a matrix row, then the "field-length" should be ten columns and can be set by using LET $T = 10 * (J - 1) + 1$. If, for example, the same matrix A is processed, the printing of its entries within fields of length 10 would result in the following:

```
1234567890123456789012345678901234567890   — labeling of columns
ROW1
1           22.33        15          44.56  ]  rows of matrix
ROW2                                        ]  A (actual output
2           -3.45        1.78        3456   ]  seen on terminal screen)
```

Remember that our goal, which may not always be possible, is to print the entries of a matrix row along one printed line. You should experiment with the instructions given in Example 2.7 on your computer to see what happens when a certain field-length is specified and the number of entries in each matrix row requires more columns for printing than the number of columns per line available for printed output.

3.1 Computations Using Matrices

The first kind of matrix computation that we shall discuss involves using computational formulas or expressions that change the values of certain entries within a matrix. This type of computation is often used to change all the values in a certain row (or column) of the matrix. An important example of this is to multiply each entry in row I of a matrix by a fixed number $N1$. Of course, values for I and $N1$ must be assigned by appropriate READ-DATA instructions or LET statements.

The fundamental type of BASIC programming instruction that we will use extensively is an assignment instruction of the form LET $A(I,K) = \underbrace{\quad}$ which will replace the current value (in computer memory) of $A(I,K)$ by the computed value of the computational expression $\underbrace{\quad}$ on the right side of the equality sign.

To illustrate this, consider the program of Example 2.3. Suppose we add the followng instructions:

```
140  LET A(2,4) = 5 * A(1,2)
142  LET A(3,1) = 2 * 5 - 38/2
144  LET A(3,4) = 10 * A(3,4)
```

These instructions will change the original matrix A given by the DATA statements in the program so that A becomes the following matrix.

$$\begin{bmatrix} 1 & 1.2 & 1.3 & 1.056 \\ 2 & 2.2 & 2.3 & 6.0 \\ -9 & 3.2 & 3.3 & 30.56 \end{bmatrix}$$

In other words, $A(2,4)$'s value of 2.056 is replaced by the value of 5 * $A(1,2)$ = 5 * 1.2 = 6.0, then $A(3,1)$'s value of 3 is replaced by 2 * 5 - 38/2 = 10 - 19 = -9, and finally $A(3,4)$'s value of 3.056 is replaced by the value of 10 * 3.056 , 30.56.

Exercise 3.1

Design BASIC instructions to perform the three calculations:

1. $A(2,3) = A(3,1) \div 15$
2. $A(1,2) = (A(1,2))^2$
3. $A(3,2) = \dfrac{3^2 - 5}{7}$

First, let us examine how we can use a computer to multiply all the entries in row I of a matrix A by a real-number which we store in a variable say $N1$. The variables representing row I are $A(I,1)$, $A(I,2)$, ..., $A(I,N)$ where N represents the number of columns in the matrix. The computations

```
LET A(I,1)  = N1 * A(I,1).
LET A(I,2)  = N1 * A(I,2)
            .
            .
            .
            .
            .
LET A(I,N) = N1 * A(I,N)
```

must all be performed. Conceptually, the computational instruction LET A(I,K) = N1 * A((I,K) must be executed for K = 1,2,...,N. This can be accomplished using a FOR-NEXT loop as follows.

```
135  FOR K = 1 TO N STEP + 1
138  LET A(I,K) = N1 * A(I,K)
140  NEXT K
```

For a complete program, we can use programming instructions similar to those used in Example 2.5 (which are designed for input and output of a matrix) and simply add the following program segment.

Example 3.1

```
131 REM MULTIPLY ROW 3 BY 5
132 LET I = 3
133 LET N1 = 5
135 FOR K = 1 TO N STEP + 1
138 LET A(I,K) = N1 * A(I,K)
140 NEXT K
```

Similar row-multiplication computations such as multiplying row 2 by $\pi \doteq 3.141$ for example can be done by changing instruction #132 to LET $I = 2$ and instruction #133 to LET $N1 = 3.141$.

Exercise 3.2

Design a program segment similar to Example 3.1 to multiply each entry in column I of a matrix A by the real number variable N1. Observe that column I is represented by

A(1,I)
A(2,I)
⋮
A(M,I)

{ and LET $A(K,I) = N1 * A(K,I)$ needs to be computed for appropriate values of K.

An important computational routine is to find the sum of the entries in any specified row of a matrix. This has many uses as the following type of information illustrates.

Household Expenses	Month 1	2	3	4	5	6
Utilities	129.34	110.24	95.74	130.45	114.32	100.47
Phone	11.66	21.35	14.12	18.23	23.35	30.54
Food	150.45	160.42	175.45	168.79	155.66	190.34
Clothes	45.35	50.43	35.67	38.55	47.87	56.43

This naturally forms a 4×6 matrix and the sum of the entries in a particular row would give the total expenses (for all six months) for the expense category (utilities, phone, etc.) associated with the row being summed. Moreover, additional matrix rows could be used to summarize data for other expense categories such as entertainment, automobile 1, automobile 2, medical, and so forth.

Similarly, data for additional months requires only additional columns.

Summing row I means to calculate $A(I,1) + A(I,2) + A(I,3) + \ldots + A(I,N)$ for some value of I. Essentially, we must program the computer to add one term at a time and successively proceed across the row entries for these terms. This process is outlined in the following flow diagram (Fig. 3.2). Here we use the variable S to compute the sum of the entries in row I by adding these values (see the summing instruction) to S one at a time starting with $A(I,K)$ where $K = 1$ and finishing with $A(I,K)$ where $K = N$.

SUMMING INSTRUCTION

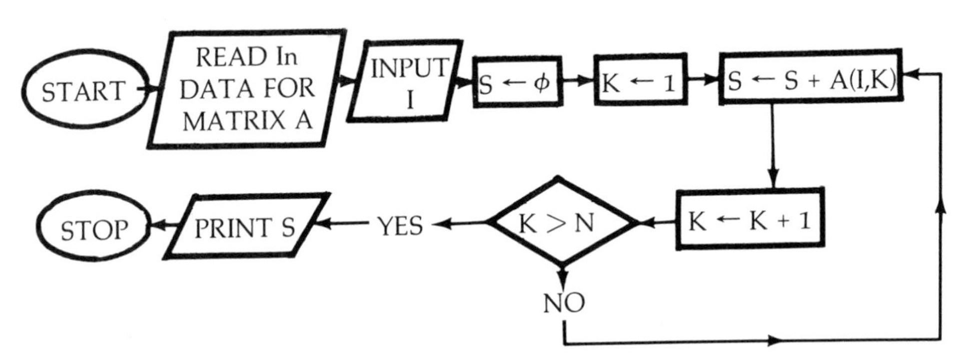

Note: ← means computer performs computation and assigns resulting value to variable on left side of ←.

Figure 3.2

Initially, the computer assigns S the value of ϕ. Then S is reassigned the value of $S + A(I,K)$ by the summing instruction for each value of K starting with $K = 1$ and ending with $K = N$. Again we have a loop process wherein the calculation $S \leftarrow S + A(I,K)$ is repeated for the values of $K = 1,2 \ldots\ldots, N$. By now, you should recognize that the computer instructions necessary for carrying out this loop process can be implemented by a FOR-NEXT loop as follows.

```
156  FOR K = 1 TO N STEP + 1
158  LET S = S + A(I,K)
160  NEXT K
```

This is the essential phase of a computer program for summing up the entries in row I. A complete program that implements the preceding flow diagram is now given. (Data for the matrix is input first, of course.)

Example 3.3

```
100  DIM A(4,6)
105  READ M
107  READ N
110  FOR I = 1 TO M STEP + 1
120  FOR J = 1 TO N STEP + 1
122  READ A(I,J)
125  NEXT J
130  NEXT I
131  REM
132  REM A MESSAGE TO USER IS PRINTED
133  REM ON HOW TO USE THE PROGRAM FOR
134  REM INTERACTIVE INPUT OF THE INDEX
135  REM OF THE ROW WHICH WILL BE SUMMED
136  REM
137  PRINT "TYPE ROW # OF ROW TO BE SUMMED"
138  PRINT "THEN PRESS RETURN KEY"
140  PRINT "OR TYPE 500 TO END PROGRAM"
141  PRINT "  "
142  INPUT I
143  REM
144  REM PROGRAM CHECKS WHETHER TEST
145  REM CONDITION, I=500, IS TRUE AND
146  REM GOES TO END WHEN TRUE OTHERWISE
147  REM PROGRAM CONTINUES WITH INSTRUC-
148  REM TIONS FOLLOWING THE IF INSTRUCTION
149  REM NAMELY LINES 152,ETC.
150  IF I = 500 THEN 400
151  REM
152  REM COMPUTE THE SUM S OF ENTRIES
153  REM IN ROW I
154  LET S = 0
156  FOR K = 1 TO N STEP + 1
158  LET S = S + A(I,K)
160  NEXT K
162  PRINT "SUM OF ROW ";I;" IS ";S
164  REM
165  REM PROGRAM IS RETURNED TO LINE
166  REM NUMBER 142 FOR REPETITION
167  REM OF PROGRAM TO FIND THE SUM
168  REM OF OTHER ROWS IF DESIRED
170  GOTO 142
190  REM DATA FOR NUMBER OF ROWS
191  DATA 4
192  REM DATA FOR NUMBER OF COLUMNS
193  DATA 6
199  REM DATA FOR MATRIX IS ARRANGED IN ROWS
200  DATA 129.34,110.24,95.74,130.45,114.32,100.47
202  DATA 11.66,21.35,14.12,18.23,23.35,30.54
204  DATA 150.45,160.42,175.45,168.79,155.66,190.34
206  DATA 45.35,50.43,35.67,38.55,47.87,56.43
400  END
```

Since this program would be used to calculate the sum of the entries of row *I* for possibly several different values of *I*, we use a computer programming method of providing data values for *I* while the program is running. This is called *interactive input*. The BASIC instruction INPUT *I* accomplishes this by printing a question mark on the screen and assigning the number typed by a user in response to the question mark. It is best to print out some information to the user as to what input is being requested and so forth whenever an INPUT instruction is used in a program. This is done by instructions 137, 138, and 140 in the program of Example 3.3. Extensive remark (REM) statements explain more fully how this program works.

The following figure shows the interactive use of this program and the resulting output.

```
]RUN
TYPE ROW # OF ROW TO BE SUMMED
THEN PRESS RETURN KEY
TYPE 5ØØ TO END PROGRAM
?1
SUM OF ROW 1 IS 680.56
?2
SUM OF ROW 2 IS 119.25
?3
SUM OF ROW 3 IS 1001.11
?4
SUM OF ROW 4 IS 274.3
?5ØØ
```

Figure 3.4

Exercise 3.3

Type in the preceding program at a computer terminal. Then run this program for different values of I and check the results. To save time, the REM statements, which are explanatory statements, may be omitted. You may use data from your household expenses if you desire. (See Exercise 2.4 for a review of how to use a computer, if necessary.)

For many applications, matrices are an important way of tabulating data in an organized manner. Various applications in accounting, for example, require calculating the sum of the entries of a given row or column. Even though we merely discussed a method for finding the sum of the entries in a specified row, this method can easily be modified so that the sum of the entries in a specified column can be calculated. In thinking about this process, remember that column I is

A(1,I)
A(2,I)
⋮
A(M,I)

and thus terms of the form $A(K,I)$ must be added to the variable S for $K = 1$ through M.

Exercise 3.4

Write a program similar to Example 3.3 that inputs a value of I and calculates the sum of the entires in column I for any M by N matrix A. Type your program in on the computer and test whether it works or not. Finally, if a printer is available, make a hardcopy listing of your program and the results of some test runs. **Note:** If the data for matrix of household expenses is used, then the sum of the entries in any given column would represent the total expenses for the associated month.

3.2 Rounding Off Matrix Entries to a Specified Precision

Before we continue our discussion of matrix arithmetic operations, we need to discuss how numbers can be rounded off to a specified accuracy (within the limitations imposed by the type of computer being used). This is important because if the original data input for the matrix is precise only to the nearest 1/100, 1/1000, etc., the output should be of the same precision. Numbers which have "mathematically speaking" an infinite decimal form like 2/3, π, etc. can only be stored using a finite number of decimal places in a computer anyway. Of course, each computer will have its own built-in method of rounding off such numbers. For example, the Apple II+ will treat 2/3 as .66666667 when this number is printed as output.

BASIC has several built-in functions like TAN( ), SIN( ), SQR ( ), etc. The function we will employ is INT ( ). INT ( ) gives the greatest integer which is less than or equal to the numerical value of ( ). For example, INT (5.786) is 5 because 5 is greater than all other integers that are less than or equal to 5.786 while INT (–6.78) is –7 because –7 is the largest of all integers which are less than or equal to –6.78.

The formula LET X = INT (100 * X + .5)/100.0 rounds the contents of X to the nearest hundredth. As an illustration, suppose X is initially 545.3678 then 100 * X = 54536.78, 100 * X + .5 = 54537.28, INT (100 * X + .5) = 54537, and finally INT (100 * X + .5)/100.0 becomes 545.37. Similarly, the formula LET X = INT (1000 * X + .5)/1000.0 rounds the contents of X to the nearest thousandth.

To round off all the entries of an M by N matrix A to the nearest thousandth, i.e., 3 decimal places of precision, a program segment consisting of instructions such as:

Example 3.5

```
FOR I TO M STEP + 1
FOR J = 1 to N STEP + 1
LET A(I,J) = INT(1000 * A(I,J) + .5)/1000.0
NEXT J
NEXT I
```

can be inserted into a program wherever it is necessary. Usually, this is done just before any set of programming instructions for printing out a matrix whose values have been changed in any way by computer calculations.

Exercise 3.5

Design a set of instructions for rounding off all entries of an M by N matrix A to the nearest 1/10000, i.e., 5 decimal places of precision.

3.3 Computer Implementation of Matrix Arithmetic Operations

The three fundamental operations of matrix arithmetic are:

(1) multiplication of a matrix by a scalar (fixed real number)

(2) addition of two matrices

(3) multiplication of two matrices

Since the entries of matrices may be numbers like 21.0135467 and matrices may have many rows and columns, it is often highly desirable to use a computer to store the matrix entries and perform accurately the tedious calculations that may be involved with any of these matrix arithmetic operations.

The multiplication of all entries in a matrix by a fixed real number is essentially an extension of the method of multiplying row I of a matrix by a real-number variable $N1$; this was discussed in Section 3.1 (see Example 3.1). We must instruct the computer to perform the calculation,

```
LET A(I,K) = N1 * A(I,K)
```

for all values of the row and column indices I and K; for an M by N matrix, this means $I = 1,2,\ldots,M$ and $K = 1,2,\ldots,N$. This can be done using nested FOR-NEXT loops. The programming instructions using this procedure (see instructions numbered 135-145) form the primary program segment in the following complete program.

Example 3.6 Program for Matrix Scalar Multiplication, $A = N1 * A$

```
100  REM PROGRAM MULTIPLIES MATRIX A
101  REM BY A SCALAR VARIABLE N1
102  DIM A(3,4)
104  REM READ IN DATA FOR MATRIX A
105  READ M
107  READ N
110  FOR I = 1 TO M STEP + 1
120  FOR J = 1 TO N STEP + 1
122  READ A(I,J)
125  NEXT J
127  NEXT I
130  REM MULTIPLY MATRIX A BY N1
131  REM ASSIGN N1'S VALUE FIRST
132  LET N1 = 5
134  FOR I = 1 TO M STEP + 1
135  FOR K = 1 TO N STEP + 1
138  LET A(I,K) = N1 * A(I,K)
140  NEXT K
145  NEXT I
150  FOR I = 1 TO M STEP + 1
151  PRINT "ROW";I
152  FOR J = 1 TO N STEP + 1
154  PRINT A(I,J)
156  NEXT J
157  PRINT "  "
158  PRINT "  "
160  NEXT I
190  REM DATA FOR NUMBER OF ROWS
191  REM AND NUMBER OF COLUMNS
192  DATA 3
193  DATA 4
198  REM DATA FOR MATRIX IS
199  REM ARRANGED IN ROWS
200  DATA 1,2,3,4
```

```
202  DATA 2,0,1,0
204  DATA 3,1,1,1
300  END
```

Exercise 3.6

In the program of Example 3.6, certain instructions must be changed to use this program to compute

$$3.5 * \begin{bmatrix} 10 & 20 & 30 & 40 \\ 5 & 6.7 & 9.6 & 100 \end{bmatrix}$$

Show which instructions must be changed and the necessary changes.

We now consider the design of a BASIC program for computing $C = A + B$ where A, B, and C are all $M \times N$ matrices. Matrix addition is defined so that each entry of C is the sum of the two corresponding entries of A and B. In other words, the calculation

C(I,J) = A(I,J) + B(I,J)

must be performed for $I = 1,2,\ldots,M$ and $J = 1,2,\ldots,N$. Again, this can be accomplished using nested FOR-NEXT loops as shown by instructions numbered 141-148 in the following complete program.

In any program, instead of a BASIC statement such as FOR $I = 1$ TO M STEP + 1, we can use the abbreviated version FOR $I = 1$ TO M. Other phases of the program are explained by appropriate comment statements, i.e., REM statements.

Example 3.7 Program for Matrix Addition, $C = A + B$

```
100  REM PROGRAM COMPUTES THE MATRIX SUM C = A + B
101  DIM A(3,4)
102  DIM B(3,4)
103  DIM C(3,4)
105  READ M,N
106  REM DATA FOR M AND N
107  DATA 3,4
112  REM
114  REM READ IN MATRIX A
115  FOR I = 1 TO M
116  FOR J = 1 TO N
117  READ A(I,J)
118  NEXT J
119  NEXT I
128  REM
129  REM READ IN MATRIX B
130  FOR I = 1 TO M
```

```
131  FOR J = 1 TO N
132  READ B(I,J)
133  NEXT J
134  NEXT I
138  REM
140  REM COMPUTE C = A + B
141  FOR I = 1 TO M
142  FOR J = 1 TO N
144  LET C(I,J) = A(I,J) + B(I,J)
145  NEXT J
147  NEXT I
148  REM
149  REM PRINT OUT MATRIX C
150  FOR I = 1 TO M STEP + 1
151  PRINT "ROW";I
152  FOR J = 1 TO N STEP + 1
154  PRINT C(I,J),
156  NEXT J
157  PRINT "  "
158  PRINT "  "
160  NEXT I
200  REM
201  REM DATA FOR MATRIX A
202  DATA 1,30,4,7
203  DATA 2,4,0,7
204  DATA 3,0,1,1
300  REM
301  REM DATA FOR MATRIX B
302  DATA 1,0,3,20
303  DATA 2,3,4,0
304  DATA 3,5,7,1
400  END
```

Exercise 3.7

In the program of Example 3.7, certain instructions must be changed to use this program to compute

$$\begin{bmatrix} 1 & 2 \\ 0 & 4 \\ 5 & 6 \\ -1 & 8 \end{bmatrix} + \begin{bmatrix} 3 & 7 \\ 5 & 0 \\ 1 & 2 \\ 0 & 4 \end{bmatrix}$$

Show which instructions must be changed and the necessary changes.

Exercise 3.8

Revise the program of Example 3.7 so that the sum of three matrices can be computed. This can be done using either one of the following methods.

Method I. To compute the matrix sum = A + B + C, use a set of nested loops to compute D(I,J) = A(I,J) + B(I,J) + C(I,J) for all values of I and J.

Method II. First compute C1 = A + B as in Example 3.7, then extend this program to incorporate additional instructions that will compute D = C1 + C.

Note: All matrices used must have DIM statements at the start of the program.

Next, we consider the design of a BASIC program to compute the product matrix $C = A * B$ where A is an $M \times N$ matrix and B is an $N \times L$ matrix. By definition, in matrix C, each entry $C(I,J)$ is computed by the following formula.

Formula 3.8

$C(I,J) = A(I,1) * B(1,J) + A(I,2) * B(2,J) + \ldots + A(I,N) * B(N,J)$. This is sometimes called the "vector dot-product" of row I of matrix A with column J of matrix B.

Although this formula looks complicated, it is a summing process and we can use the following set of BASIC programming instructions for this computation (compare with Example 3.3).

```
144  LET S = 0
145  FOR K = 1 to N
146  LET S = S + A(I,K) * B(K,J)
147  NEXT K
149  LET C(I,J) = S
```

Observe that with this FOR-NEXT loop, the instruction LET S = S + A(I,K) * B(K,J) will compute the sum of the terms $A(I,1) * B(1,J)$, $A(I,2) * B(2,J)$, ..., $A(I,N) * B(N,J)$ as this instruction is repeated for the values of $K = 1,2,\ldots,N$. Thus, $C(I,J)$ will be calculated according to Formula 3.8.

Now, to find the product matrix C, we must calculate $C(I,J)$ using Formula 3.8 for all values of the indices I and J, i.e., $I = 1,2,\ldots,M$ and $J = 1,2,\ldots,L$. In effect, we need to instruct the computer to repeat the calculations in the above program segment for $I = 1,2,\ldots,M$ and $J = 1,2,\ldots,L$. This can be accomplished by including this program segment inside a set of nested loops which make I and J take on the desired values. Example 3.8 gives a complete program for computing the matrix product and makes use of this approach.

Example 3.8 Program for Matrix Multiplication, $C = A * B$

```
100  DIM A(3,4)
101  DIM B(4,2)
102  DIM C(3,2)
103  REM
104  REM READ IN MATRIX PARAMETERS
105  READ M,N,L
106  REM DATA FOR M,N,L
107  DATA 3,4,2
113  REM
114  REM READ IN M BY N MATRIX A
115  FOR I = 1 TO M
116  FOR J = 1 TO N
117  READ A(I,J)
118  NEXT J
119  NEXT I
128  REM
129  REM READ IN N BY L MATRIX B
130  FOR I = 1 TO N
131  FOR J = 1 TO L
132  READ B(I,J)
133  NEXT J
134  NEXT I
139  REM
140  REM COMPUTE M BY L MATRIX C = A * B
141  FOR I = 1 TO M
142  FOR J = 1 TO L
143  REM COMPUTE THE SUM: A(I,1) * B(1,J) + . . .  + A(I,N) * B(N,J)
144  LET S = 0
145  FOR K = 1 TO N
146  LET S = S + A(I,K) * B(K,J)
147  NEXT K
149  LET C(I,J) = S
150  NEXT J
151  NEXT I
167  REM
168  REM PRINT OUT MATRIX C
169  PRINT "THE MATRIX PRODUCT IS"
170  FOR I = 1 TO M
171  PRINT "ROW"; I
172  FOR J = 1 TO L
173  PRINT C(I,J),
174  NEXT J
175  PRINT "  "
176  PRINT "  "
177  NEXT I
199  REM
200  REM DATA FOR MATRIX A
201  DATA 1,30,4,7
202  DATA 2,4,0,1
203  DATA 3,0,1,1
299  REM
300  REM DATA FOR MATRIX B
301  DATA 1,0
```

```
302 DATA 0,1
303 DATA 0,0
304 DATA 1,1
400 END
```

This Example Computes:

$$C = \begin{bmatrix} 1 & 30 & 4 & 7 \\ 2 & 4 & 0 & 1 \\ 3 & 0 & 1 & 1 \end{bmatrix} * \begin{bmatrix} 1 & 0 \\ 0 & 1 \\ 0 & 0 \\ 1 & 1 \end{bmatrix}$$

Exercise 3.9

Indicate the changes in the program of Example 3.8 which must be made in order to compute

$$C = \begin{bmatrix} 0 & 2 & 3 & 0 & 1 \\ 1 & 0 & 0 & 2 & -2 \end{bmatrix} * \begin{bmatrix} 5 & 2 \\ 0 & 1 \\ 1 & 0 \\ 0 & -1 \\ 1 & 0 \end{bmatrix}$$

3.4 Application of Computer—Matrix—Multiplication to Dietary Analysis

A computer program for matrix multiplication is useful for finding the product of matrices with several rows and columns since the computer can perform the lengthy computations very rapidly and without making arithmetic errors. Another important reason is that extensive numerical information may be stored as part of a computer program (using DATA statements) for one or more of the matrices being used in the computations. This allows for repeated use of a table of fixed data values when performing matrix computations based on such information.

An excellent illustration of this is to use matrices to compute various parameters of an individual's diet based on the kinds of foods and amounts consumed during a day or week. A simplified example of this is based on the following table of food composition (the foods shown in this table were selected according to the author's dietary preferences).

Table 3.9[1]

Food Item	Measure	Calories	Carbo-hydrate g	Protein g	Fiber g	Fat g	Choles-terol mg	Sodium mg
Ground beef (lean)	1 lb.	812	0	94	0	45	295	0
Chicken (thigh)	1 lb	435	0	62	0	34	368	377
Snapper (filet)	1 lb.	422	0	90	0	5	0	304
Liver (chicken)	1 lb.	585	13	89	0	20	2517	318
Salmon (pink, canned)	1 cup	310	0	45	0	13	77	0
Peanuts (roasted)	1 cup	838	30	38	4	70	0	18
Yogurt (plain)	8 oz.	144	16	12	0	3.5	14	159
Banana	1 avg.	127	33	1.6	.8	.3	0	2
Apple (raw)	1 med.	96	24	.3	1.8	.1	0	2
Butter	1 tbsp.	102	.1	.1	0	12	35	140
Cheese (Cheddar)	1 oz.	112	.4	7	0	9.4	30	176
Egg, large	1	82	.5	6.5	0	6.4	312	61
Whole wheat bread	1 slice	56	11	2.4	.4	.7	0	121
Bran flakes	1 cup	106	28	3.6	1	.6	0	207
Mayonnaise	1 tbsp.	101	.3	.2	0	11.2	10	84
Apple juice	1 cup	117	30	.2	.3	0	0	2
Milk (low fat)	1 cup	121	12	8.1	0	4.7	18	122

[1]Sources: Various Food Labels and *Nutrition Almanac* by Nutrition Search, Inc. (See Bibliography.)

We can modify the program of Example 3.8 and store the information of Table 3.9 in a 17×7 matrix B where the rows of B consist of the data about calories, carbohydrates, . . . , sodium for the 17 associated food items. The information on measure is not included in B but serves to indicate the units to be used in keeping records on the amounts of these foods that are consumed during a specified time interval.

Now assume, for example, the following amounts of these foods are consumed during a week.

Figure 3.10

Food	Amount	Food	Amount
Ground beef	2.5 lb.	Butter	21 tbsp.
Chicken	3.5 lb.	Cheese	3 oz.
Snapper	1.5 lb	Eggs	5
Liver	0 lb.	Whole wheat bread	20 slices
Salmon	1 cup	Bran flakes	3 cups
Peanuts	2 cups	Mayonnaise	8 tbsps.
Yogurt	24 oz.	Apple juice	15 cups
Banana	7	Milk	7 cups
Apple	7		

Consider the 1 × 17 matrix A consisting of the numbers representing the amounts of Fig. 3.10 so that

$$A = [\ 2.5 \quad 3.5 \quad 1.5 \quad 0 \quad 1 \quad 2 \quad 24 \quad 7 \quad 7 \quad 21 \quad 3 \quad 5 \quad 20 \quad 3 \quad 8 \quad 15 \quad 7\ ]$$

We can use a computer to calculate $C = A * B$ and obtain the 1 × 7 matrix

$$C = [\ 18924.5 \quad 1689.2 \quad 1185 \quad 467 \quad 935.6 \quad 5029.5 \quad 14025.5\ ].$$

We interpret the resulting useful information as follows:

$C(1,1) =$ 18924.5 represents the total calories consumed
$C(1,2) =$ 1689.2 represents the total carbohydrates (g) consumed
$C(1,3) =$ 1185 represents the total protein (g) consumed
$C(1,4) =$ 41.7 represents the total fiber (g) consumed
$C(1,5) =$ 935.6 represents the total fat (g) consumed
$C(1,6) =$ 5029.5 represents the total cholesterol (mg) consumed
$C(1,7) =$ 14025.5 represents the total sodium (mg)

This model of dietary analysis can be expanded by including more matrix rows with the same kind of information for additional food items. Also, additional columns for nutritional information on vitamin A, vitamin B_1, folic acid, potassium, etc., i.e., various vitamins, minerals, and essential amino acids could be included for each food item. The use of computers to store and make calculations with such extensive information (as in Table 3.9) would be almost mandatory. Moreover, the programming instructions and data for the tabular information can be stored in computer memory, floppy disks, etc., and loaded back into the computer very rapidly. Then, the information for the matrices can be updated repeatedly and analyzed by computer processing. In the model of dietary analysis this would usually mean updating (on a daily or weekly basis) the information on food consumption used for the entries of matrix A; of course, additional data on relevant food items in an individual's diet could also be stored by adding additional rows to matrix B.

Exercise 3.10

Suppose we have the following additional information:

Food Item	Measure	Calories	Carbo-hydrate	Protein	Fiber	Fat	Choles-terol	Sodium
Bisquick	.5 cups	240	38	4	0	8	0	700
Buttermilk								
Flour								
Rice (brown) (cooked with salt)	1 cup	178	38	3.8	15	1.2	0	423
Turkey (cooked) (dark meat)	1 lb	921	0	136	0	38	458	449

Use the above information and Table 3.9 or information from food labels to set up a matrix for food composition of staple items in your own diet. Modify the program of Example 3.8 to process information relevant to your own weekly food consumption. If a printer is available, obtain a hard copy of a listing of your program and the resulting output.

4. Computer Implementation of Elementary Row Transformations on Matrices

In the Gaussian Elimination Algorithm, there are three row operations that can be used to transform a given matrix into a row-equivalent matrix. These are as follows:

(1) Interchange any two rows of a matrix,

(2) Divide a given matrix row by a fixed, nonzero real number,

(3) Add a multiple of one matrix row to another matrix row.

Let us consider first the elementary row operation of interchanging row I with row J. Suppose matrix A has N columns so row I is $A(I,1)\ A(I,2)\ldots A(I,N)$ and row J is $A(J,1)\ A(J,2)\ldots A(J,N)$. To carry out this interchange, $A(J,1)$ is replaced by $A(I,1)$, $A(J,2)$ is replaced by $A(I,2)$, etc. Then $A(I, 1)$ must be replaced by original value of $A(I,1)$ — not the new value of $A(J,1)$, which after replacement has become that of $A(I,1)$. Similarly, $A(I,2)$ must be replaced by the original value of $A(J,2)$ and so on.

Since each entry of one row is replaced by the entry in the same column of the other row, the computer instructions to do these reassignments will essentially have the same form. Thus, we use a computer looping technique to perform this row interchange process. The following program segment accomplishes this.

Example 4.1

```
230 FOR K = 1 TO N STEP + 1
240        LET T = A(J,K)
250        LET A(J,K) = A(I,K)
260        LET A(I,K) = T
270 NEXT K
```

This FOR NEXT loop, with K the loop variable, makes each instruction inside the loop (lines 240, 250, 260) can be executed for each value of K. Thus, initially $K = 1$, T is assigned the value of $A(J,1)$, then $A(J,1)$ is assigned the value of $A(I,1)$, and finally $A(I,1)$ is assigned the value of T. Next $K = 2$, T is assigned the value of $A(J,2)$, then $A(J,2)$ is assigned the value of $A(I,2)$, and finally $A(I,2)$ is assigned the value of T. This repetitive procedure continues for subsequent values of K and ends with $K = N$.

Here T is a temporary variable and stores the original value of $A(J,K)$ for each K. T is necessary because $A(J,K)$ will be replaced by the value of $A(I,K)$, so thereafter the value of T is used to replace the value of $A(I,K)$ with the original value of $A(J,K)$. This can be described schematically as follows:

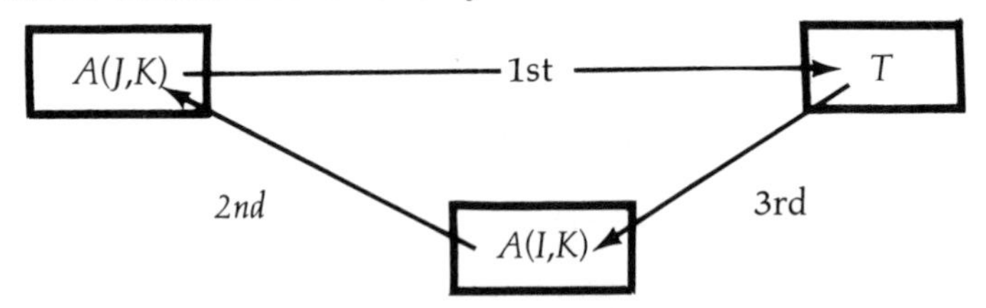

Let us now consider the design of complete program (or subprogram which could be used as part of a more comprehensive program) to carry out the row-interchange process using the method discussed in Example 4.1. Since the choice of the two rows that need to be interchanged must usually be made while the program is running on a computer, we need to make use of the idea of interactive input for the values of I and J. This can be accomplished by a BASIC instruction of the form INPUT I, J. When the computer executes this instruction, values for I and J must be supplied by the user. A complete program with explanations (REM statements) is shown by Example 4.2.

Example 4.2 Program To Interchange Two Rows

```
100 DIM A(3,4)
104 REM READ IN # OF ROWS AND
105 REM # OF COLUMNS
107 READ M,N
108 REM READ IN DATA FOR MATRIX A
110 FOR I = 1 TO M STEP + 1
120 FOR J = 1 TO N STEP + 1
122 READ A(I,J)
125 NEXT J
130 NEXT I
135 REM LOOP TO ROUND-OFF ENTRIES
136 REM AND PRINT OUT MATRIX A
140 PRINT "  "
148 PRINT "  "
150 FOR I = 1 TO M STEP + 1
151 PRINT "ROW"; I
152 FOR J = 1 TO N STEP + 1
153 LET A(I,J) = INT (1000 * A(I,J) + .5) / 1000.0
154 LET L = 7 * (J - 1) + 1
155 PRINT TAB ( L); A(I,J);
156 NEXT J
157 PRINT "  "
158 PRINT "  "
160 NEXT I
200 PRINT "TYPE #'S OF ROWS TO BE"
201 PRINT "INTERCHANGED OR TYPE  Ø, Ø"
202 PRINT "TO END RUNNING OF PROGRAM"
211 REM INPUT VALUES FOR I AND J
212 REM AND TEST IF I = Ø is TRUE
213 REM TO END PROGRAM
215 INPUT I,J
220 IF I = Ø THEN 600
225 REM LOOP FOR INTERCHANGE
230 FOR K = 1 TO N STEP + 1
240 LET T = A(J,K)
250 LET A(J,K) = A(I,K)
260 LET A(I,K) = T
270 NEXT K
280 GOTO 140
500 REM DATA FOR M AND N
501 DATA 3,4
510 REM DATA FOR ROWS OF MATRIX A
511 DATA 1,2,3,4
512 DATA 2,8,-3,-1
513 DATA 3,0,8,7
600 END
```

Exercise 4.1

(i) Sign on at a computer terminal and follow the procedures for your computer system so that you can use BASIC. Then type in the above program segment. See Exercise 2.4

if you need guidance on how to type a program on a computer. The REM statements can be omitted, of course. Finally, by typing the system command RUN, have the computer execute your program for several values of I,J.

Note: The instructions 200-202 are PRINT commands and the phrases enclosed by the quotation marks will be printed out when the program is run. Next, instruction 215 asks for values of I,J to be typed in after the question mark. These numbers must be separated by a comma when typed and will be relayed to the computer by pressing the return key. To terminate further repetitions of this row-interchange process, you must type Ø, Ø in response to the program's request for values for I,J.

(ii) Make the necessary adjustments in the DATA and DIM instructions so that the program processes

$$A = \begin{bmatrix} -1 & .5 & 7 & 9 & .1 \\ 8 & 3 & 6 & 0 & .2 \\ -5 & 0 & 0 & .1 & .3 \end{bmatrix}$$

A good program will work for matrices in general and require only changes in the DATA and DIM statements for different matrices.

The next exercise is designed to further your understanding of the computer instructions for implementing matrix transformations. Suppose we want the computer to interchange column I with column J rather than rows. Using the computer notation for columns in matrix A, we would have

$$\begin{bmatrix} \mathrm{A(1,I)} \\ \mathrm{A(2,I)} \\ \vdots \\ \mathrm{A(M,I)} \end{bmatrix} \text{ and } \begin{bmatrix} \mathrm{A(1,J)} \\ \mathrm{A(2,J)} \\ \vdots \\ \mathrm{A(M,J)} \end{bmatrix} \text{ for columns } I \text{ and } J.$$

Now for a column interchange $A(1,J)$ must be temporarily stored and then replaced by $A(1,J)$ and afterwards $A(1,I)$ must be replaced by the stored value of $A(1,J)$. Similarly $A(2,J)$ must be temporarily stored and then replaced by $A(2,J)$ and afterwards $A(2,I)$ must be replaced by the stored value of $A(2,J)$. This process must be continued for all entries in these columns. Thus, the instruction will be very similar to those for the row-interchange program and a

```
For K = 1 to M STEP + 1
  .
  .
  .
NEXT K
```

loop should be used. Here M (a variable) represents the number of rows in the matrix and is, therefore, the number of entries in each column. Thus the loop variable K will range through the indices of the column entries. This method should be used to solve the following exercise.

Exercise 4.2

Write a program similar to Example 4.2 which reads in a matrix A and values of I,J as input and then interchanges column I with column J. Type your program in on the computer and test whether it works or does not. Finally, make a hard-copy listing of this program and the results of some test runs if a printer is available.

Let us now consider the elementary row operation of dividing each entry in row I by a fixed nonzero number represented by the variable D. An understanding of the computer techniques for implementing this type of row operation should in some ways be easier than those required for the row-interchange operation. Here row I, which in general is $A(I,1)\ A(I,2)\ A(I,3)\ldots A(I,N)$, will become $A(I,1)/D\ A(I,2)/D\ A(I,3)/D\ldots A(I,N)/D$ where N is the number of columns in matrix A.

In other words each entry in row I is replaced by the result of that entry being divided by D. Specifically, we will have $A(I,1)$ replaced by $A(I,1)/D$, $A(I,2)$ replaced by $A(I,2)/D$, etc. Using a more abstract notation, we see that $A(I,K)$ is replaced by $A(I,K)/D$ where K ranges through the indices $1,2,3,\ldots,N$ of the row entries. The computer instructions to implement this process should now be obvious. The actual instructions are given in the following program segment.

Example 4.3

```
320  PRINT "INPUT ROW NUMBER AND NONZERO
     VALUE OF DIVISOR"
325  INPUT I,D
330  FOR K = 1 TO N STEP +1
340       LET A(I,K) = A(I,K)/D
350  NEXT K
```

Here we are assuming the matrix A is already in computer memory, and we have given only those computer instructions in BASIC that are designed specifically for carrying out the row operation of dividing each entry of row I by a nonzero number D. Note that the values of I and D will be necessary input when the program is run and line 325 is executed by the computer.

Suppose we wanted a complete program similar to Example 4.2 that would read in a matrix A, input values of I,D and then divide row I by D. Thus, in addition to the preceding program segment, we would need several commands for reading in matrix A and so on.

Exercise 4.3

Design a complete program similar to Example 4.2 to read in a matrix A and divide row I by a nonzero number D where the values of I and D can be supplied interactively. Use an IF statement to test the condition $I = \emptyset$ and branch to the END statement when $I = \emptyset$ is true.

Suppose instead of dividing each entry in row I by a nonzero number D, you wanted to design a program to divide each entry in column I by a nonzero number D. As in our thinking about the row operation of this type, we look at the notational representation of this process. Column I is

$$\begin{bmatrix} A(1,I) \\ A(2,I) \\ \vdots \\ A(M,I) \end{bmatrix}$$

if there are M rows in matrix A as before. We have $A(1,I)$ replaced by $A(1,I)/D$, $A(2,I)$ replaced by $A(2,I)/D$, and so on with $A(M,I)$ replaced by $A(M,I)/D$ to complete this process. As before, this can be implemented by a FOR $K = 1$ to M STEP + 1, NEXT K loop. Use this method to solve:

Exercise 4.4

Write a *program-segment* like Example 4.3 which assumes the computer has values for a matrix A and is given the values of I,D as input and then divides each entry of column I by D.

The third type of row operation is that of adding a multiple of row I to row J and thereby creating a new row J. Here each entry of row I is multiplied by a number, which can be represented by the variable R, and this result is added to the corresponding entry of row J; finally this sum replaces the existing value of that entry in row J. Using computer notations, we have $A(J,1)$ replaced by $A(J,1) + R * A(I,1)$, $A(J,2)$ replaced by $A(J,2) + R * A(I,2)$, and so on until the last entry in row J, i.e., $A(J,N)$ is replaced by $A(J,N) + R * A(I,N)$. Thus, we can see that, in general, $A(J,K)$ is replaced by $A(J,K) + R * A(I,K)$ where K ranges through the indices 1,2,3,... N of the row entries. The computer instructions to implement this row operation are given as lines 410, 430, 440 in the following program segment. This also includes an INPUT instruction which gives the computer values for I, J, R by means of interactive input.

Example 4.4 Routine to Add: Multiple of Row I to Row J

```
400  PRINT "INPUT ROW #, ROW # BEING ADDED TO, #
     MULTIPLYING BY"
420  INPUT I,J,R
425  FOR K = 1 TO N STEP + 1
430  LET A(J,K) = A(J,K) + R * A(I,K)
440  NEXT K
```

Suppose $A = \begin{bmatrix} 1 & 2 & 3 & 4 \\ 5 & 6 & 7 & 8 \\ 2 & 3 & 6 & 7 \end{bmatrix}$, to replace row 2 of A by row 2 + (–5) * row 1 when this program is RUN, we need only type 1,2,–5 and press the RETURN key in response to the ? typed by the computer. Note row 2 would become (0 –4 –8 –12), and for the matrix given, this could be the first step in changing it to echelon form.

Exercise 4.5

(i) What numbers must be given to I,J,R to replace row 3 of A (above) by row 3 + (–2) * row 1?

(ii) What is the resulting row 3 that would be obtained?

Exercise 4.6

Write a *program-segment* like Example 4.4 (which assumes values are stored for a matrix A) to input values for I,J,R interactively and then replace each entry of column J by the sum of that entry of column J with the result of multiplying the corresponding entry of column I by a number R. In our notation, $A(K,J)$ is replaced by $A(K,J) + R * A(K,I)$ where K ranges through the indices 1,2,3,... of the column entries. Type your program in on the computer and test whether or not it works. Finally, make a hard-copy listing of this program and the results of some test runs.

A more general approach to this third type of row operations is to replace Row J by adding a multiple of Row J, say $R2$ * (Row J) to a multiple of Row I, say $R1$ * (Row I). Specifically, we would have each $A(J,K)$ replaced by $R1 * A(I,K) + R2 * A(J,K)$ when K ranges through the indices $1,2,3,\ldots,N$ of the row entries.

This is often very useful in creating a 1 as the first nonzero element of Row J without introducing fractional numbers. Consider the matrix $A = \begin{bmatrix} 7 & 12 & 3 & 1 \\ 5 & 4 & 2 & 0 \\ 10 & 3 & 7 & 4 \end{bmatrix}$. To create a 1 in the A (1,1) location, it is impossible to avoid introducing fractions when using only the limited technique of replacing Row 1 by the result of dividing either Row 1, Row 2, or Row 3 by 7, 5, or 10 respectively. We can, however, replace Row 1 by multiplying Row 1 by −2 and add this to Row 2 multiplied by +3. This gives Row 1 = (−2) * (Row 1) + (+3) * (Row 2) = (−2) * (7 12 3 1) + (+3) * (5 4 2 0) = (1 −12 0 −2).[1]

This new version of Row 1 will be much easier to use in subsequent steps of the Gaussian-Elimination Method and the propagation of fractions in the matrix can be avoided or postponed until the last stages of the Elimination Method. In fact, if the original matrix contains only integers[2], the introduction of fractions during the computational steps of the Elimination Method can be postponed usually until the very end. This is accomplished by judicious employment of this more general row operation to create a "1" as the first nonzero entry of row (a "leading 1") which would then be used as a "pivot," i.e., to create 0's in the column entries below this "leading 1."

Thus, we need to revise Example 4.4 to implement this more general type of row operation. This can be done by simply changing the instructions at lines 400, 405, and 430 so that these become

```
400  PRINT "TYPE: VALUES FOR I,J (ROW # BEING
     REPLACED), R1, R2 (#'s TIMES ROWS I,J)"
420  INPUT I,J,R1, R2
430  LET A(J,K) = R2*A(J,K) + R1*A(I,K)
```

To use this program on $A = \begin{bmatrix} 7 & 12 & 3 & 1 \\ 5 & 4 & 2 & 0 \\ 10 & 3 & 7 & 4 \end{bmatrix}$ and

[1]This was possible because the numbers 7 and 5 are relatively prime.

[2]If the original matrix has any entries which are proper fractions, each row can be multiplied by the greatest common denominator (of all the denominators in that row) to create a matrix with integer entries only.

replace Row 1 by (–2) * (Row 1) + (+3) * (Row 2), as previously discussed, would involve typing 2,1,+3,–2 in response to the computer's request for input values for *I*,*J*, *R*1, *R*2 when the program is RUN.

Finally, we observe that the previous version of this row operation can be implemented by this more general computer program if we recognize that the instruction *A*(*J*,*K*) = *A*(*J*,*K*) + *R***A*(*I*,*K*) is *A*(*J*,*K*) = *R*2**A*(*J*,*K*) + *R*1**A*(*I*,*K*) with *R*2 = 1 and *R*1 = *R*. Thus, the replacement of row *J* by adding Row *J* to a multiple (*R*1) of Row *I* can always be implemented by using 1 for *R*2.

5. Computer Implementation of the Gaussian Elimination Algorithm

We have now discussed each of the elementary row operations and developed a computer routine to implement each type of row operation. Suppose we want to use the computer to read in a matrix and then implement the algorithm for row-reducing that matrix to echelon form. We must be able to direct the computer to carry out a designated row transformation on our matrix for each step of this systematic process.

Our previous computer routines for the three types of row operations gives us three subprograms which can become parts of a more comprehensive program to carry out this algorithm. In fact, all we need is a method of telling the computer which of these three computer subprograms to carry out. This can be accomplished by attaching an identification number to each of these subprograms. Simply by inputting the number 1, 2 or 3 the computer will be directed to the subprogram which does the row operation associated with a given number.

A flow-diagram indicating this decision-making process and the overall program development is shown by Fig. 5.1.

As this flow diagram indicates by inputting a value of 1,2, or 3 for the variable X when the program is run, the computer will branch to the subprogram associated with the input value and carry out the chosen row operation and at that time request the necessary information required for this row operation. Of course, after any given row operation has been performed, the resulting matrix will be printed out. Moreover, the program flow will return to the instructions for input of X and the next choice of a new operation can be given. The program stops when a value of Ø is input for X.

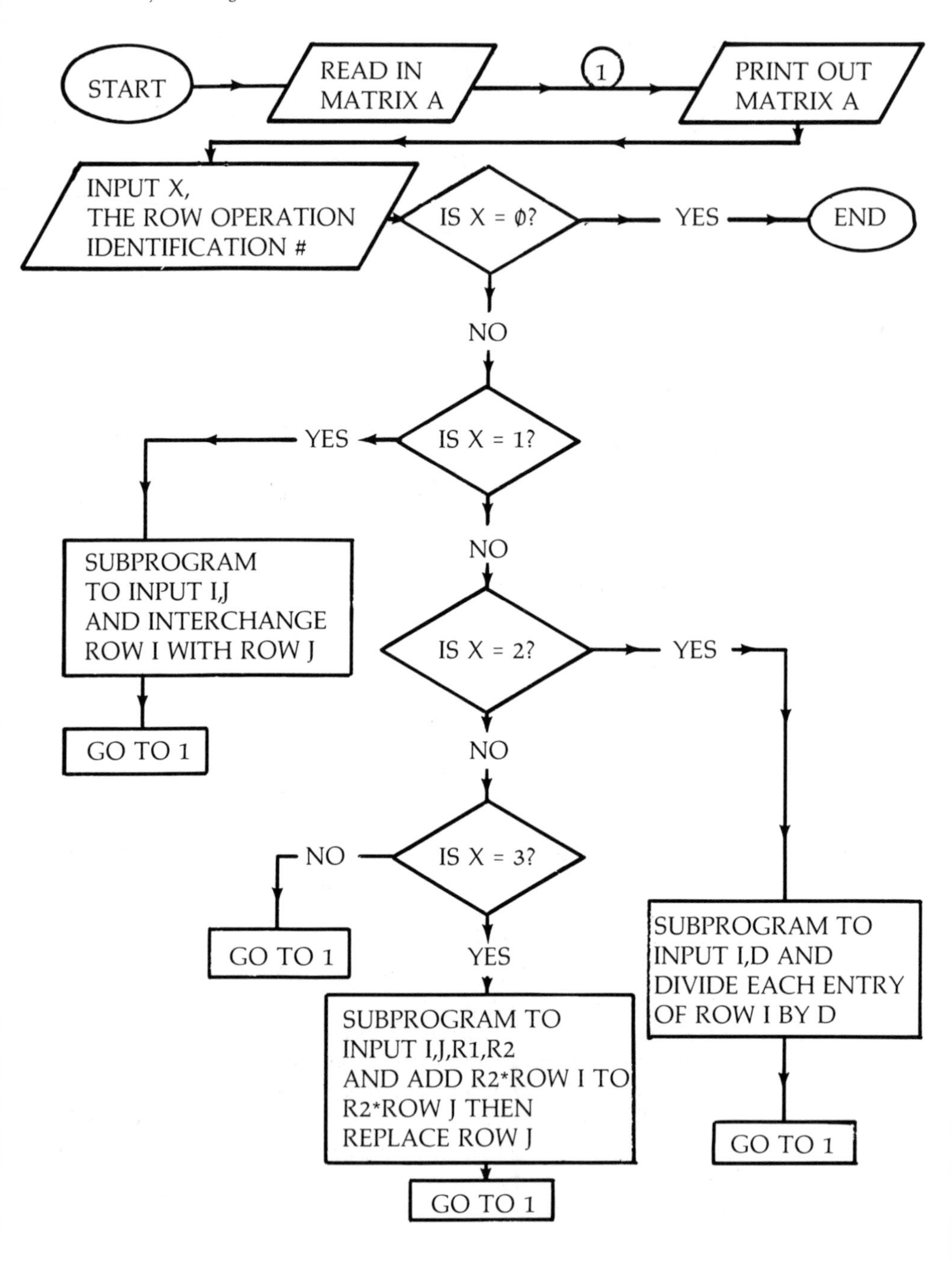

Figure 5.1

A complete program corresponding to the flowchart of Fig. 5.1 is given as follows.

Example 5.2: Program for Interactive Implementation of Elimination Algorithm

```
100  DIM A(3,4)
104  REM READ IN # OF ROWS AND
105  REM # OF COLUMNS
106  READ M,N
107  REM
108  REM READ IN DATA FOR MATRIX A
110  FOR I = 1 TO M STEP + 1
111  FOR J = 1 TO N STEP + 1
112  READ A(I,J)
114  NEXT J
115  NEXT I
139  REM
140  REM LOOP TO ROUND-OFF ENTRIES
141  REM AND PRINT OUT MATRIX A
142  REM ENTRIES ARE PRINTED IN
143  REM FIELDS OF 7 COLUMNS
145  PRINT "MATRIX IS CURRENTLY AS FOLLOWS"
148  PRINT "  "
150  FOR I = 1 TO M STEP + 1
151  PRINT "ROW"; I
152  FOR J = 1 to N STEP + 1
153  LET A (I,J) = INT (1000 * A(I,J) + .5) / 1000.0
154  LET L = 7 * (J - 1) + 1
155  PRINT TAB ( L); A(I,J);
156  NEXT J
157  PRINT "  "
158  PRINT "  "
160  NEXT I
162  IF X1 = 3 THEN 420
163  REM
164  REM SUBPROGRAM TO PRINT INFORMATION
165  REM ON HOW TO SELECT DESIRED ROW
166  REM OPERATION AND DECISION STEPS
167  REM FOR COMPUTER TO BRANCH TO
168  REM ASSOCIATED SUBPROGRAM FOR
169  REM OPERATION SELECTED
170  PRINT "INPUT # FOR DESIRED ROW OPERATION"
171  PRINT "1 IS FOR ROW INTERCHANGE"
172  PRINT "2 IS FOR DIVIDING A ROW BY A #"
173  PRINT "3 IS FOR ADDING MULTIPLES OF ROWS"
174  PRINT "0 IS TO END RUNNING OF PROGRAM"
175  PRINT "  "
177  INPUT X
178  PRINT "  "
180  IF X = 0 THEN 600
181  IF X = 1 THEN 200
182  IF X = 2 THEN 300
183  IF X = 3 THEN 400
184  PRINT "YOU TYPED AN INVALID #"
185  PRINT "  "
186  PRINT "PLEASE TRY AGAIN"
190  GOTO 140
191  PRINT "  "
```

Example 5.2: (Continued)

```
200  REM SUBPROGRAM FOR ROW INTERCHANGE
201  PRINT "INPUT #'S OF ROWS TO BE INTERCHANGED"
202  PRINT "OR TYPE 0,0 TO SELECT ANOTHER OPERATION"
203  PRINT "  "
211  REM INPUT VALUES FOR I AN J
212  PRINT "  "
215  INPUT I,J
220  IF I = 0 THEN 140
224  REM
225  REM LOOP FOR INTERCHANGE
230  FOR K = 1 TO N STEP + 1
240  LET T = A(J,K)
250  LET A(J,K) = A(I,K)
260  LET A(I,K) = T
270  NEXT K
280  GOTO 140
299  REM
300  REM SUBPROGRAM FOR ROW DIVISION
321  PRINT "INPUT ROW # AND NONZERO DIVISOR"
323  PRINT "OR TYPE 0,0 TO SELECT ANOTHER OPERATION"
324  PRINT "  "
325  INPUT I,D
327  IF I = 0 THEN 140
328  REM BRANCH TO INPUT OF I,D AGAIN IF D = 0 IS TRUE
329  IF D = 0 THEN 321
330  FOR K = 1 TO N STEP + 1
340  LET A(I,K) = A(I,K) / D
350  NEXT K
355  GOTO 140
394  REM
395  REM SUBPROGRAM FOR ADDITION OF ROW MULTIPLES
400  PRINT "INPUT VALUES FOR I,J,R1,R2 ——"
401  PRINT "J IS THE ROW BEING REPLACED,"
404  PRINT "R1 IS THE # TIMES ROW I AND R2"
405  PRINT "IS THE # TIMES ROW J"
407  PRINT "#'S MUST BE SEPARATED BY COMMAS"
408  PRINT "  "
409  PRINT "OR TYPE 0,0,0,0 FOR ANOTHER SELECTION"
410  PRINT "  "
411  REM SET X1 = 3 SO COMPUTER DOES LINE 420 AFTER PRINTING
     A, THEN X1 IS SET TO 4
413  LET X1 = 3
414  PRINT "HERE IS MATRIX A AGAIN"
415  GOTO 150
420  INPUT I,J,R1,R2
421  LET X1 = 4
422  IF I = 0 THEN 140
425  FOR K = 1 TO N STEP + 1
430  LET A(J,K) = R2 * A(J,K) + R1 * A(I,K)
440  NEXT K
445  GOTO 140
500  REM DATA FOR M AND N
501  DATA 3,4
```

```
510 REM DATA FOR ROWS OF MATRIX A
511 DATA 1,2,3,4
512 DATA 5,6,7,8
513 DATA 2,3,6,7
600 END
```

Observe that the instructions for the three subprograms are essentially those contained in example programs 4.2, 4.3, and 4.4. The key new features are instructions in lines 177-183 which request a value for X as input then use IF... THEN... instructions to: branch to line 200 when $X = 1$ is true so the subprogram for row interchange can be carried out; branch to line 300 when $X = 2$ is true so the subprogram to divide entries of ROW I by D can be carried out; branch to line 400 when $X = 3$ is true so the subprogram to replace ROW J by R1*ROW I + R2*ROW J can be carried out.

The use of this computer program is illustrated by the following results of a computer run for the matrix

$A = \begin{bmatrix} 1 & 2 & 3 & 4 \\ 5 & 6 & 7 & 8 \\ 2 & 3 & 6 & 7 \end{bmatrix}$. A copy of the computer row-reducing process is shown by Fig. 5.3.

The first row operation should be to add −5 * Row 1 to Row 2 and replace Row 2 with this. Thus, the number for desired-row operation is 3; then the numbers 1,2,−5,1 are input as the values or $I,J,R1,R2$ as required by this row operation. As shown on the copy of the printout, the resulting matrix is $\begin{bmatrix} 1 & 2 & 3 & 4 \\ 0 & -4 & -8 & -12 \\ 2 & 3 & 6 & 7 \end{bmatrix}$.

Next we should replace row 3 by adding −2 * Row 1 to Row 3. Therefore, we again desire row operation 3 and the numbers 1,3, −2,1 are input as the values necessary for this row operation. The resulting matrix printed out is then $\begin{bmatrix} 1 & 2 & 3 & 4 \\ 0 & -4 & -8 & -12 \\ 0 & -1 & 0 & -1 \end{bmatrix}$.

The remaining steps in carrying out the algorithm to reach echelon form are shown (see Fig. 5.3) on the copy of the computer printout and are sufficiently explained there.

Figure 5.3

DISPLAY OF INTERACTIVE USE — COMMENTARY

]RUN
MATRIX IS CURRENTLY AS FOLLOWS

ROW1
1 2 3 4

← *Original Matrix*

ROW2
5 6 7 8

ROW3
2 3 6 7

INPUT # FOR DESIRED ROW OPERATION
1 IS FOR ROW INTERCHANGE
2 IS FOR DIVIDING A ROW BY A #
3 IS FOR ADDING MULTIPLES OF ROWS
Ø IS TO END RUNNING OF PROGRAM

?3

INPUT VALUES FOR I,J,R1,R2 —
J IS THE ROW BEING REPLACED,
R1 IS THE # TIMES ROW I AND R2
IS THE # TIMES ROW J
#'S MUST BE SEPARATED BY COMMAS

{ *First Step of Elimination Algorithm*
Replace Row 2 by –5 * Row 1 + Row 2

OR TYPE Ø, Ø, Ø, Ø FOR ANOTHER
SELECTION

HERE IS MATRIX A AGAIN
ROW1
1 2 3 4

ROW2
5 6 7 8

ROW3
2 3 6 7

?1,2,–5,1
MATRIX IS CURRENTLY AS FOLLOWS

{ First Step Requires $I = 1$, $J = 2$,
$R1 = -5$, $R2 = 1$

ROW1
1 2 3 4

ROW2
Ø –4 –8 –12

← *Matrix Resulting from First Step*

ROW3
2 3 6 7

Fig. 5.3 (Continued)

```
INPUT # FOR DESIRED ROW OPERATION
1  IS FOR ROW INTERCHANGE
2  IS FOR DIVIDING A ROW BY A #
3  IS FOR ADDING MULTIPLES OF ROWS
Ø  IS TO END RUNNING OF PROGRAM

?3

INPUT VALUES FOR I,J,R1,R2 —
J IS THE ROW BEING REPLACED,
R1 IS THE # TIMES ROW I AND R2
IS THE # TIMES ROW J
#'S MUST BE SEPARATED BY COMMAS

OR TYPE Ø, Ø, Ø, Ø FOR ANOTHER
SELECTION

HERE IS MATRIX A AGAIN
ROW1
1    2    3    4

ROW2
Ø    -4    -8    -12

ROW3
2    3    6    7

?1,3,-2,1
MATRIX IS CURRENTLY AS FOLLOWS

ROW1
1    2    3    4

ROW2
Ø    -4    -8    -12

ROW3
Ø    -1    Ø    -1

INPUT # FOR DESIRED ROW OPERATION
1  IS FOR ROW INTERCHANGE
2  IS FOR DIVIDING A ROW BY A #
3  IS FOR ADDING MULTIPLES OF ROWS
Ø  IS TO END RUNNING OF PROGRAM

?2

INPUT ROW # AND NONZERO DIVISOR
OR TYPE Ø, Ø TO SELECT ANOTHER
OPERATION
```

{ *Second Step of Elimination Algorithm*

Replace Row 3 by -2 * Row 1 + Row 3 }

{ Second Step Requires $I = 1$, $J = 3$,

$R1 = -2$, $R2 = 1$ }

← *Matrix Resulting From Second Step*

{ *Third Step of Elimination Algorithm*

Divide Row 2 by -4
Row # is 2, Divisor is -4 }

Fig. 5.3 (Continued)

```
?2,-4
MATRIX IS CURRENTLY AS FOLLOWS

ROW 1
1    2    3    4

ROW2
Ø    1    2    3                  ←Matrix Resulting From Third Step

ROW3
Ø    -1    Ø    -1

INPUT # FOR DESIRED ROW OPERATION
1  IS FOR ROW INTERCHANGE
2  IS FOR DIVIDING A ROW BY A #
3  IS FOR ADDING MULTIPLES OF ROWS
Ø  IS TO END RUNNING OF PROGRAM

?3

INPUT VALUES FOR I,J,R1,R2 —
J IS THE ROW BEING REPLACED,
R1 IS THE # TIMES ROW I AND R2
IS THE # TIMES ROW J
#'S MUST BE SEPARATED BY COMMAS

OR TYPE Ø, Ø, Ø, Ø FOR ANOTHER    { Fourth Step of Elimination Algorithm
SELECTION                         { Replace Row 3 by 1 * Row 2 + Row 3

HERE IS MATRIX A AGAIN
ROW1
1    2    3    4

ROW2
Ø    1    2    3

ROW3
Ø    -1    Ø    -1
                                  { Fourth Step Requires I = 2, J = 3,
? 2,3,1,1                         { R1 = 1, R2 = 1
MATRIX IS CURRENTLY AS FOLLOWS

ROW1
1    2    3    4

ROW2
Ø    1    2    -3                 ← Matrix Resulting From Fourth Step

ROW3
Ø    Ø    2    2
```

```
INPUT # FOR DESIRED ROW OPERATION
1  IS FOR ROW INTERCHANGE
2  IS FOR DIVIDING A ROW BY A #
3  IS FOR ADDING MULTIPLES OF ROWS
Ø  IS TO END RUNNING OF PROGRAM

?2

INPUT ROW # AND NONZERO DIVISOR
OR TYPE Ø, Ø TO SELECT ANOTHER
OPERATION
?3,2
MATRIX IS CURRENTLY AS FOLLOWS

ROW1
1    2    3    4

ROW 2
Ø    1    2    3

ROW3
Ø    Ø    1    1

INPUT # FOR DESIRED ROW OPERATION
1  IS FOR ROW INTERCHANGE
2  IS FOR DIVIDING A ROW BY A #
3  IS FOR ADDING MULTIPLES OF ROWS
Ø  IS TO END RUNNING OF PROGRAM
```

{ *Final Step of Elimination Algorithm:* Divide Row 3 by 2 }

{ Row # is 3, Divisor is 2 }

←*Row Echelon Form of Original Matrix*

***** Ø Is Typed to End Running of Program *****

Exercise 5.1

Sign on at a computer terminal and use appropriate procedures for your system to call up the MATRIX program, which should have been created and saved by you or your instructor previously. Make appropriate changes in the data statements so that a matrix of your choice (perhaps some augmented matrices for problems in your textbook which involve systems of linear equations) can be read in when the program is run. Then run the program and use it to row-reduce your matrix to echelon form. Obtain a hard copy of your procedures and the sequence of resulting matrices.

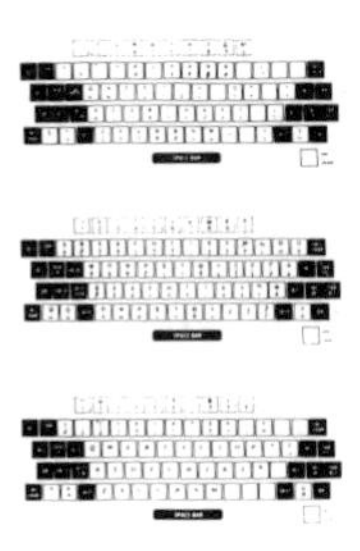

6. Bibliography

Atkinson, K. E., *An Introduction to Numerical Analysis,* John Wiley & Sons, Inc., New York, NY, 1978.

Arden, B. and K. Astill, *Numerical Algorithms: Origins and Applications,* Addison-Wesley, Reading, MA, 1970.

Beckett, R. and J. Hurt, *Numerical Calculations and Algorithms,* McGraw-Hill, New York, 1967.

Bent, R. J. and G. Sethares, *BASIC: An Introduction to Computer Programming,* Brooks/Cole Publishing Company, Monterey, CA, 1982.

Conte, S. D. and C. de Boor, *Elementary Numerical Analysis,* McGraw-Hill, New York, NY, 1980.

Dorn, W. S., G. G. Bitter, and D. L. Hector, *Computer Applications for Calculus,* Prindle, Weber and Schmidt, Inc., Boston, MA, 1972.

Forsythe, G.E., "Pit Falls in Computation, or Why a Math Book Isn't Enough," AMM 77 (1970), 931-956.

Forsythe, G. E. and C. Moler, *Computer Solution of Linear Algebraic Systems,* Prentice-Hall, Inc., Englewood Cliffs, NJ, 1967.

Joues, R. M., *Introduction to Computer Applications Using BASIC,* Allyn and Bacon, Inc., Boston, MA, 1981.

Mott, J. L., A. Kendell, and T. Baker, *Discrete Mathematics for Computer Scientists,* Reston Publishing Company, Reston, VA, 1983.

Motter, W. L., *Elementary Techniques of Numerical Integration and Their Computer Implementation,* UMAP Module U379, Education Development Center, Inc., Newton, MA, 1982.

Moursund, D. and C. Duris, *Elementary Theory and Application of Numerical Analysis,* McGraw-Hill, New York, NY, 1967.

Nutrition Search, Inc., *Nutrition Almanac,* McGraw-Hill, New York, NY, 1979.

Olinick, M., *An Introduction to Mathematical Models in the Social and Life Sciences,* Addison-Wesley, Reading, MA 1978.

Ralston, A., *A First Course in Numerical Analysis,* McGraw-Hill, New York, NY, 1965.

Roberts, F. S., *Discrete Mathematical Models,* Prentice-Hall, Inc. Englewood Cliffs, NJ, 1976.

Steinberg, D. I., *Computational Matrix Algebra,* McGraw-Hill, Inc., New York, NY, 1974.

Wilkinson, J. H., *Rounding Errors in Algebraic Processes,* Prentice-Hall, Inc., Englewood Cliffs, NJ, 1963.

Williams, G. *Mathematics With Applications in the Management, Natural, and Social Sciences,* Allyn and Bacon, Inc., Boston, MA, 1981.

Williams, G., *Computational Linear Algebra With Models,* Allyn and Bacon, Inc., Boston, MA, 1982.

7. Answers to Selected Exercises

Exercise 2.1 (possible answer)

```
100  DIM A(2,6)
120  READ A(1,1),A(1,2),A(1,3)A(1,4)
121  READ A(1,5),A(1,6)
122  READ A(2,1),A(2,2),A(2,3)A(2,4)
123  READ A(2,5),A(2,6)
200  DATA 3.46,2,5,7.01,5,6.3
202  DATA 77.5,3.6,4,9,100.1,8.7
300  END
```

Exercise 2.2 (possible answer)

```
100  DIM A(3,5)
105  READ M
110  FOR I = 1 TO M STEP +1
120  READ A(I,1),A(I,2),A(I,3)A(I,4),A(I,5)
130  NEXT I
190  REM DATA FOR # OF ROWS
191  DATA 3
199  REM DATA FOR ROWS OF MATRIX
200  DATA 1,7,8,6.2,90
202  DATA 2,5.03,7,17,10
204  DATA 3,66.7,50.5,3,70
300  END
```

Exercise 2.3

Example 1.3's program can be used with the following modification: 100 DIM A(4,3); 191 DATA 4; 193 DATA 3; 200 DATA 20, 1.2, 7.5; 202 DATA 15,3,66.7; 204 DATA 8,5,5; 206 DATA 9.9,1,6.0

Exercise 3.1

```
1. LET A(2,3) = A(3,1)/15
2. LET A(1,2) = A1,2) * A(1,2)
3. LET A(3,2) = (3 * 3 - 5)/7
```